THE REVOLUTIONS OF TIME

By

JONATHAN DUNN

The Revolutions Of Time
by Jonathan Dunn

ISBN: 978-93-59327-07-5

Published by

DOUBLE 9 BOOKS
2/13-B, Ansari Road, Daryaganj
New Delhi – 110002
info@double9books.com
www.double9books.com
Tel. 011-40042856

ABOUT THE AUTHOR

Jonathan Dunn, best known for his literary masterwork "The Revolutions of Time," is a gifted writer whose passion is evident in his work. His extraordinary capacity to integrate historical analysis with human experience distinguishes him in the world of writing. Dunn's work acts as a bridge, bringing people together and encouraging understanding and empathy. Dunn's ingenuity and enthusiasm are obvious inside the pages of his writings. His stories transcend the everyday, transporting readers on enthralling journeys through various realms of emotion and intellect. Dunn's writing is distinguished by its infinite originality and zeal, whether it is analyzing historical events or probing the depths of human psychology. Jonathan Dunn's publications are distinguished by their exquisite yet approachable prose. He has a special talent for making difficult concepts and sophisticated stories accessible to all. His extraordinary stories attest to his storytelling talent, allowing readers from all walks of life to relish in the complexity of his narratives. In conclusion, Jonathan Dunn's literary legacy is a woven tapestry of knowledge, imagination, and passion. His ability to combine historical study with the human condition and express it in an elegant, approachable manner guarantees that his books continue to resonate with readers from all walks of life, making him a beloved figure in the literary world.

CONTENTS

Chapter 1
Past and Present...8

Chapter 2
Predestined Deja Vu...11

Chapter 3
Zards and Canitaurs...14

Chapter 4
Onan, Lord of the Past 23

Chapter 5
The Treeway..30

Chapter 6
The Fiery Lake..35

Chapter 7
Down to Nunami...47

Chapter 8
The Temple of Time..55

Chapter 9
Mutually Assured Deception..62

Chapter 10
Devolution...68

Chapter 11
The Land Across the Sea..78

Chapter 12
The White Eagle..90

Chapter 13
The Big Bang 97

Chapter 14
Past and Future...103

...The very men who claimed mental superiority because they were free from superstitions and divine disillusionment were themselves victims of their own sophism, and while they thought themselves crowned with enlightenment, it was naught but the Phrygian caps of their prejudices toward the material state.

- Jehu, the Kinsman Redeemer

The physical manifestation of the spiritual force is not the spiritual force at all, only a bland deception. If you only focus on what you can see directly, than you chase after only the representation and not the object desired. If a bird is flying through the sky at noontime, casting a shadow on the ground below him, and a man comes along, and in the hope of catching the bird chases after its shadow, it is evident that he will never catch it, for when he does reach it, he will find that there is nothing there at all, only the shadow of what it was he desired. So it is with the spiritual!

- Onan, Lord of the Past

Chapter 1
Past and Present

My name is Jehu. Most probably it sounds foreign and unfamiliar to you, devoid of the qualities of affection and personality which give character to a name. It is a harsh name, cold and inhuman, like something out of the night, an unwelcome intruder into the warmth of familiarity. It inspires no blissful memories, nor does it kindle fond feelings in the bosom of the hearer, instead the heart is hardened to it like the feathers of a duck to water, repulsing it, leaving it to run off into the ditches and by-ways of the long forgotten past, to trickle dejectedly into those stagnant ponds where so many words of wisdom are imprisoned: out of sight, out of mind, out of heart, out of history. Yet while history is forgotten and misconstrued, it is repeated, for what is life without water, which nourishes and sustains it, and what is life without wisdom, which protects and cultivates it?

Jehu is my name, though it no longer brings the quickened pulse and keen anticipation of happiness to the hearts of any, not even my own. For what deference can be given to a name, though not in itself a thing of dishonor, which represents the failure to derail the evitable fate which wrecks the race of man again and again. Not that I myself embody such a failure, nor even that I gave birth to the dreaded fate's latest momentum, but as is seen time and again throughout history, one name is brought to represent the tide of change, for better or worse, the doer of deeds which were done not by him, but by a mass of independent doers, yet it is written in the annals of history as the deeds of but one man.

While I had little to do, consciously, with the doom of the earth, I will always be fingered as the villain, as the ambitious Napoleon or the barbaric Atilla, the arrogant Augustus or the fearful Cyrus. Someone has to bear the burden of shame on the pages of history for the people of his time, and in that sense, maybe I truly can be called their kinsman redeemer. Perhaps it is my fate to bear witness to the wrongs of a people, of which even you are not wholly innocent.

And yet can an individual be blamed for the faults of a society, can personal responsibility be extended to the members of an unknown multitude? How the enjoined conscience of one longs to say no, but in good faith it cannot be said, for in this case the mask of ignorance cannot supersede the face of guilt. Indeed, ignorance in this case only adds to the shame of the guilty, this being a crime not of misdeeds but of negligence, twisted together with the vices of humanity into a thick and sturdy cord, a rope that cannot be pulled apart and individually examined, yet must be taken as a whole. Insularly, the strand of ignorance could be easily snapped, remedied by but a little education, yet when woven together by one's own hands with prides and prejudices, it forms an unbreakable rope, which is placed about our neck to hang us: through means of our own doing is our fate foretold. If but one or two of the strands were omitted, the result would be a feeble rope, easily broken, and we would live. But by our own vices is our mortality made manifest, by our own wrongs are we wronged.

By now you may be beginning to feel the impulses of indignation arising in your breast, for who am I, the admittedly despicable Jehu, to group you as my fellow convicts, my co-conspirators, in a sense? And you are right, for I am not your judge and neither do I wish to be.

Having said that, I now request of you to put down the book and discontinue reading.

"Surely," you say to yourself, "He is mentally deranged, for what author in his right mind would encourage his readers to disperse, what writer does not thrive on the digestion of his words by an eager audience?"

Here I must make a revelation to you: if my manuscript has indeed been found, then I have long since been dead; and I assure you that in whatever form my existence takes in the present, I have little desire for your intrigue or goodwill. Do you think Melville is consoled in death of his miserable life by the vainglorious praises of the living? Or do you think that Poe is comforted by such avid attentions in his present abode? In truth, Melville's only rivalry is now within, and Poe's only raven that daunting memory of those truths which had escaped him in life, but which now are opened to you.

More importantly, if this manuscript has been found, it proves that what is contained herein is the unerring truth. I do not write this to exonerate myself, however let me say here that I am more the Andre' than the Arnold, for I was but the emissary of history, not the traitor to humanity, and if not me then some other would have filled the void. Let it be remembered that it was Andre' who gave his life for his deeds, and yet it is Andre' who is recollected with a sweet sorrow, and though Arnold lived, he had no peace. Yet while history is vivid and encyclopedic, in itself a living organism, it can speak only through the mouths of men, who often misrepresent it for their own partisan and prejudiced plans. It is strong and steadfast, though, and in time is always victorious over its menial opposition, for what is history but the past tense of truth, and it is justly said that *veritas numquam perit*, truth never dies.

Going back to what I said before, namely that at my manuscript's discovery my demise will itself be history: I am assured that such is true, for even now as I write this my death is near at hand. How wide the abyss of time that separates us is I cannot tell, but I do know that it is beyond the reckoning of men, such an unknown barrage of hollow, formless years. Yet as you read this it is as if I were speaking directly to you, despite all of the desolation between our times. That is what makes history an organic being, and by history I mean all of the past, or all of the future, depending on your viewpoint.

A book is a connection between times and peoples, more so than any other medium. As I put these words down in writing, it is as if I am imparting my very self into the pages. And as you read them, the name Jehu slowly forms into an image, into a personality, and from the empty word Jehu comes the great well of affection springing from a personal intimacy. A book is an enigma in which no time exists, and as it is read it brings the reader into its eternal being, for while it sits closed on a shelf it is no more than a forgotten memory, yet when it is opened its contents come to life and its characters and locations are once more existent in the same state as when they were written, the story becomes once more reality.

While I have long been deceased, when you read this I am brought to life once more, and with my rebirth I tell you my story, and make known to you the truths contained therein. The words of this book are a rune gate, a portal to the past, and as you read them, your present fades away and you are drawn into my present, this very moment in which I now write. Then you connect with me intimately, and for a brief time the gulf of mortality is transcended and the depths of my being are laid open to you. We commune together and you eat of my flesh and drink of my blood, merging your existence with mine.

Come to me now, my friend, come to me across the gulf of mortality, for I await you. Come, and in your spiritual peregrination meet with me, in this land of the past which is so foreign and unfamiliar to you, but which will become for a time your home. Come to me, my friend, and let me tell you my story.

Chapter 2
Predestined Deja Vu

It was in the last stages of sleep that I began to feel the warm morning sun strike my face, and hear the pleasant chirping of birds and crickets. I rolled slowly over, stretched my legs and my back, and stood up, with the last remnants of a dream playing quietly in my mind. But as I came to my feet and got a clear view of where I was, I realized it was not a dream that I had had at all, but something far more sobering. I found myself somewhere in the center of a very large prairie which covered the land for many miles around. From the sun's lowly position on the eastern horizon, it was evident to me that the new day was just dawning, casting a golden hue on the grasses that covered the prairie's surface.

Around the distant outskirts of the plain I could make out a ring of trees circumventing the whole, waving almost imperceptibly to and fro in the light breeze that was blowing. A few miles to the southwest there was a group of odd looking trees stretching up over the horizon to a considerable height. They were closer than the outer ring, which kept a uniform girth around the prairie, but somehow they looked very peculiar and foreboding, and I got one of those sobering feelings which I like to call predestined deja vu. What I mean is that I got a sense of deja vu, but instead of the past converging with the present into one thought, the present seemed to converge with the future, and the result was a mysterious foreboding of something, though I couldn't tell what. That is the sensation that I had when I saw what I assumed to be a small grouping of trees somewhere in the southwestern portion of the savanna, though that was merely a guess, for in the distance I could only make out several dark forms rising out of the grassland like trees, or possibly buildings, one of them being a great deal taller than the others, with a spherical shape on top that only faintly resembled a tree's crown. If it was indeed a tree, it was the largest that I have ever seen, for it looked to be upwards of 800 feet tall.

My mental warning bells were ringing quite loudly, and I endeavored to silence them by extreme exertions of the will, but they would not be subdued. I assumed that they were not at all correct, much like the fearful expectancy some have while swimming in the ocean, out of sight of all land, of being attacked by an enormous leviathan of the deep. As unfounded as the fear is, it places one into a frenzy of dubious thoughts that inspire equally frantic and anarchist actions. Because of this, I thought that my ideas were naught but superstitious fancies, yet try as I might, I could not rid myself of them.

Instead, I made up my mind to set off in the opposite direction, north, and to advance at a double march until I should reach the woody border, which looked to present shelter not only from the southern apparitions, but also from the shielded underworld of the grasses, in which also dwelt the mysterious sense of fear and predestined deja vu. It was slightly chilly, but beyond that nothing defaced the temperate beauty of the day, and even that promised to soon dissipate with the continual strengthening of the sun's warmth. As I walked, or rather, trotted along, it did just that, and in the growing warmth of the day the sweet fragrances of the many various grasses rose to the surface, delighting my odor perceiving sensors with their earthy simplicity.

The day marched on, and with it I, and the distant wall of trees began to slowly grow closer. At length, I found myself at their edge, at around the noon hour, and as I came upon the first of them, I leaned against the trunk of a large, thickset tree for a moment of repose and reflection in its shade. It was by all appearances an ancient wood, for the line between it and the prairie was distinct, appearing as if the shrubs and lesser flora had acquiesced to fate and retreated beyond the forest's claimed boundaries, rather than continue for countless ages to charge and then be pushed back, to gain a foothold only to be thrown out a year or two later. The trees themselves were mighty pinions of strength, tall and of great girth, and spread far apart from one another, leaving wide open spaces between their towering trunks. A short, soft grass clothed the land that stretched on in their midst, joined in its solitude by a hearty looking moss that stretched itself out on the trunks of the trees and on the rocks and boulders that lay scattered here and there among the open spaces. Far above, the trees' great branches spread out a thick canopy, covering the whole of the forest area in a relaxing and invigorating twilight, rendering itself homely and quaint. After a few moments of enjoying that most pleasing scene, I roused and extricated myself unwillingly from its enchanted depths and set off once more into the heart of the woods, having no where else to go.

After a time, I cannot say how long, I came upon a small, trickling stream which flowed deeper into the woods, that direction being northward. A short walk along its path, after refreshing myself to content with its pure waters, brought me to its destination: a large lake into which the forest opened. Its banks were very gradual and the grass of the woodland led right up to the water's edge. The surface of the water itself was smooth and delicate.

Amidst the pleasantness of the scene, there was something missing from the feel of the area: inhabitants. There was an abundance of wild life of all kinds, and much organic life as well, but something greater than flora or fauna was missing: people. I had traveled so far, and without any sighting of a person. It was a lonely and desolate feeling which prevailed, despite the abundances of life. Novelties soon grow worthless with no one to share them with, ideas become meaningless if not communicated timely, emotions grow boisterous and uncontrollable with no end to receive them.

I was quite alone, unfortunately, and it dampened my spirits considerably. Feeling despondent, I turned and walked sullenly from the lake's edge into the woodland once more, with no definite purpose in mind, only a meandering thought of my dismal situation. My thoughts morphed, in succession, from anxiety to despair, to anger, to frustration, and in my frustration I knelt down and picked up a fallen branch from the ground, walked to the nearest tree, and eyed a strange, protruding knob that stuck out from the trunk. I held the branch at shoulder's length and swung it at the knob with all the force of my built up emotions. It hit with a crash and a hollow thud, leaving the branch broken and my arm sore, but the knob undamaged.

But then something unexpected happened: with a grating noise, a small hole appeared part way up the trunk, coming from what looked to be solid wood, for no sign was seen before of its having an opening. From the newly opened hole was then thrust out a head, hairy and with a short snout-like edifice for a nose and mouth. Its eyes and the furry hair which covered its face were brown, and a few wily whiskers protruded from its snout. With a look of utter surprise, as if it had not expected me as much as I had not expected it, it eyed me closely for a moment and then looked anxiously from side to side and told me to come in.

When those words passed its lips, or whatever artifice it spoke from, a great weight fell from my shoulders. After a short moment, quickened by my relief, a door appeared in the trunk of the tree, its edges previously hidden behind the thick mosses. Swinging inwards, it opened and revealed the creature standing there, beckoning me to enter. I did, and the door shut behind me, leaving me in the darkness of the hollow tree.

Chapter 3
Zards and Canitaurs

My eyes quickly adjusted to the darkness, and once they did I saw that the trunk was hollowed out to the extent of eight feet in diameter, with two stairways, one up and another down, filling either corner of the small entry room in which I found myself. Observing that my vision was returned enough to see, the strange creature which had greeted me led me down the descending staircase for a short way, until we came into a cavern which was delved beneath the roots of the tree.

The walls and floor of the cavern, or more accurately, the sitting room, for such it appeared to be, were paneled with a thick, heavy wood with an almost artificially symmetric grain, and the ceiling was done in diagonal boards of the same. Sitting in the center of the room was a brick-laid pit in which burned an illuminating fire, and around it was placed an odd covering frame that caught up the smoke and channeled it via underground passages to some distant wilderness, where its sightless remnants would dissipate into the atmosphere unnoticed. On the near side of the fire was a round table flanked by four large, comfortable chairs, padded by cushions made from the same material as the various carpets and tapestries around the room.

There were two more of the strange creatures seated at the table, called Canitaurs as I later found out, and as they are closely entwined with my story, being prominent participants, I will describe them in some detail here. They stood erect like a man, yet were quite contrasted in appearance. Their skin for one was covered in a thick, impenetrable coat of hair, much like a dog or a bear's. Their hands, also, were less distinct in the fingers, though but slightly, and their limbs were a little longer and thicker than a man's. The two most notable differences, however, were the formation of their shoulders and chest, which were very pronounced and muscular, and their faces. The latter's features were brought to a point in the short snout, or muzzle, that formed their nose and mouth, taking their chins with it and leaving a long line from their neck to their chest open. Humanity prevailed in the rest of their features, though, giving them the look of a man and canine hybrid.

By then I had overcome my initial perplexion at the sight of the Canitaurs, and I endeavored to put a strong check over my emotions in order to prevent another outbreak of panic and to remain cool and candid, come what would. Yet it was, ironically, the product of my rashness that I had found their habitation at all. This I successfully did, and as I entered the room, led by the Canitaur who was on watch, the others stood politely and greeted me with an apparent intrigue.

Our conversation proceeded at follows:

"I am Wagner of the Canitaurs, my friend," said the one who appeared to be the leader, "And these are Taurus and Bernibus," the latter being the one who had led me down. "Welcome to Daem."

"I am Jehu," I told them, "It is a pleasure to meet you."

"Indeed, and under such circumstances as well. Tell me, how did you come to be here?"

Here I smiled nervously, and replied, "I am a traveler from a distant land, and came here by the advice of a friend."

At this somewhat false answer, more in character than in content, Wagner looked at me wonderingly, as if detecting my falsehood, but did not follow his look with any probing questions, to my great relief. In order to steer the conversation away from this point, I added quickly, "I am not at all disappointed, either, for the landscape is beautiful and the trees and foliage are wondrously large, but I was surprised to find that, from the prairie to the lake, I saw no one living among these quaint locations."

Wagner looked at me closely, with a hint of almost reverencing respect and said, "You were very fortunate in your travels, I assure you, for had you arrived at any other time, you would have fallen into fouler hands than ours by far."

"I do not understand what you mean," I said.

"Of course not, I am forgetting your new arrival has left you unacquainted with affairs that I am faced with everyday. Let me explain: we, that is, the Canitaurs, have been in open hostilities with the other group of people on this island, the Zards, for as long as we can remember. They have great military superiority in this section of Daem, and when we come here we are forced to live in hiding, in outposts such as this one."

"Why not just make peace?" I asked.

"Because it is our ideologies that conflict, neither group of us will yield, and the solution can only be decided by force, military force. It is fortunate that you have come among us first, for they would have mistreated you."

"So you have said, though I do not see why I was not captured by them on my journey through the plains, if they are as powerful in this quarter as you say," I replied.

"As I said, the timing of your arrival was very fortunate," he said, "At any other time you would have surely been caught, and then your fate would have been uncertain, but yesterday was the Zard's new year, the Kootch Patah, on which they spend all night in celebrations and revelries. Because of this, they were all soundly asleep on your trip through the prairie, very possibly laying at your feet, covered by the tall grasses."

So my fears were not as unfounded as I had thought, was my predestined deja vu, then, real as well? Only time would tell.

"I am indeed lucky then, as you have said, not only in the Zard's unattentiveness, but also in finding of your secreted habitation, as well as your friendly welcoming of me," I said.

"I must confess," he chuckled, "It is not merely from a one-sided hospitality that you are welcomed."

"Indeed?" I said.

"Indeed," he answered, "For your appearance and the circumstances of your arrival are almost uncannily the realizations of one of our most ancient prophesies, one which we have longed to have fulfilled."

"Is that so?" I rhetorically asked.

"Surely it is," he said with a smile, though from happiness or humor I could not tell. He went on soberly, saying: "The prophecy is concerning the kinsman redeemer, one of the ancients sent by Onan, the Lord of the Past, to redeem us from the destruction of this polluted world."

"What do you mean by 'one of the ancients'?" I interjected questioningly.

"Exactly what I said," Wagner replied with a light hearted smile, "Let me explain."

But before he could, we were interrupted by a violent scratching and pounding at the door, along with some grunting voices which I could not understand. The Canitaur's ears, which were quite large, though more erect and postured than floppy, quickly rose to attention, and they had spent not a moment listening when they uniformly chorused, "Zards," in a hoarse whisper. My earlier fear, then mysterious but now understood, returned in full force, and my face writhed in horror as I ejaculated remorsely, "Then we are lost."

Wagner turned gravely towards me and said, "Perhaps, but there is still hope. Come, follow me," and rising from his chair he led the way to the furthest corner of the room. A primitive tapestry was hanging there, and

Wagner lifted it up while Bernibus and Taurus hit two hidden switches, one being on either extremity of the room, to avoid discovery. That unlocked the wall behind the tapestry. It opened along lines previously concealed by the wood's grain and revealed a small cubbyhole built into the wall, probably meant for its present use, concealment. Wagner led us into it and no sooner was the door, or wall, latched again than the Zards, having broken down the outside door by brute strength, flooded into the room.

We could see them as they did, for the wall that concealed us had many small holes, and the tapestry as well, so that on the inside we could see all that happened in the well lit room, while they could not see us, as there was no light to reveal us. Indeed, I had been sitting facing the hidden compartment during our brief dialog and had not detected it at all. The situation was quite different at that time, though, for the Zards were actively looking for us, whereas I was merely glancing occasionally at the wall.

Now that they were closer, I could easily understand their conversation:

"Blast it, they aren't here," said one,

"Probably deserted the place after Garlop saw them, he should have kept watch."

"Why? He couldn't have stopped a group of them, and they're too keen to be followed."

"Aye, he did right to hurry off, but it would be a shame if they escaped," another joined.

"The King is here though, and there's no fooling him."

"Hear ye, hear ye," the others assented, that being a common phrase among them which was the equivalent of an 'I agree' or 'Amen'.

A larger, more commanding Zard, whom the others looked in deference to, then came down the stairs, saying as he entered the room, "Let us not celebrate prematurely, gentlemen. There is nothing of interest above, so we will have to search carefully down here."

"Sir, is it true it was a hairless one he saw?" one asked him.

"We are all hairless here," he said, laughing with the others, "But yes, it is reported that Garlop saw one of the ancients, and with his sharp eyes and knowledge of history, it is assumed to be true. I need not remind you, then, the need to find them before they are too far away, it is imperative to the cause that the ancient is not brought to the hidden fortress of our adversaries."

The Zards then set to work with great assiduity searching for any clues of the Canitaur's whereabouts, examining everything meticulously, yet quickly. They tore the furniture apart to look for hidden compartments.

followed the smoke pipes through the ground to their outlets, tore off the floor boards to look for secret passages, and did the same to the ceiling.

Before I continue with my story, let me pause for a moment to describe to you the appearance of the Zards, for you are probably curious as to what they look like.

Quite different from the Canitaurs, they were, in fact, completely hairless, being almost lizard-like. They stood erect, about the same height as a man, that is, about six feet or a little over that, and their bodies resembled those of alligators, with short, thickset legs, stout arms, and a long body with a tail draping down to the ground, looking like a giant tongue, though covered, of course, in scales. Their heads were small, having a little skull on which were the eyes and ears and with a long snout that, like the Canitaurs', held their noses, mouths, and chin. Huge, sharp teeth filled their mouths and gave them an odd, fiercely sophisticated look. Their hands were thick with long fingers, and though their overall appearance had an air of awkwardness about it, they set to their tasks with great dexterity, though if it was natural or the result of their excited state, I could not tell. Indeed, I began to grow worried when the Zard who was removing the walls, to check for holes or tunnels, drew near to us as he methodically pried off the panels with a metal bar and looked for anything suspicious.

He moved along quickly and was just about to put the bar to our covering and pull when another Zard, on the other end of the room, held aloft a piece of paper, calling the attentions of the others to it. Our almost discoverer went himself to the other Zard, and we were, for a moment at least, saved from being exposed. Having read the paper, the taller Zard, the King, said to the others, "Well done, lads. We have here a map to the Canitaur's hidden fortress. Let us go to Nunami, gather some troops, and surprise them. Today may prove victorious, so let us hurry."

The others assented and as a body they went up the stairs and out the door, hurrying forth, it seemed, to do their dastardly deeds, and in their ardor not leaving behind even a single one to guard the hideout. Despite our good fortunes, my spirits were damp, for my sorrow of the Canitaur's ill fate was as a wound in my bosom, knowing that I had been the sole reason for their discovery. What a good kinsman redeemer, I thought, for my coming may have ended the wars, or put its completion in motion, yet not in the favor of my hosts.

To my chagrin, however, the Canitaurs, led by Wagner, were buxom, seeming to find great humor in what had happened. Turning to them in a zealous perplexity, I said spiritedly, "How can you laugh? You may have escaped, but your brethren are doomed, and you yourselves will not last long around enemies without the protection of the other Canitaurs."

But my rebuke only seemed to make their laughter and mirth more hearty, and they raged on without ceasing for a time. After a while, when they were reduced to a smiling remnant of their former pleasure, Wagner turned gravely towards me and said, "Forgive me, Jehu, for not explaining it to you. You are right to chastise us, but the situation is not as you seem to think it, for the map they found was a fake, and will lead them to nowhere of importance, while we affect our escape. We are lucky that they left no guard, but come, let us not tempt fate and remain any longer in this compromised outpost, to the fortress we go!"

He finished and met with the approbations of the others, and accordingly, we exited the cubby hole and made our way through the rummaged room, up the stairs, and out of the tree. It was now early evening, and the temperance of twilight, with its soft and mellow splendors, only increased the pleasantness of the area. A slight breeze prevailed and rustled the leaves and boughs of the giant trees just enough to render it pacifying and comforting. Being quickened by the breeze, the lake danced on in its earlier smoothness, only in a faster tempo, improving the ruggedness of the watery wrinkles. The last visiting rays from the sun were congregated on the eastern shores, saying their good-byes to the glowing trees, and giving their parting respects before being whisked away to their native lands of fire, to come again in great numbers on the morrow.

We set off around the lake, making our way northward towards the rugged mountains rising before us in a grand show of might. Wagner and Taurus walked before and behind us, respectively, Wagner leading the way and Taurus erasing the marks of our passing, and both watching for any signs of ambush. Bernibus walked abreast of myself, keeping me in pleasant company, for he was a very enjoyable companion.

During our walk, Bernibus and I had an insightful conversation, of which I will relate to you the following, as you may find it interesting:

"Tell me," I said to him, "You seem to be a jovial people, despite the war that you find yourselves in, but are all of your people of the same attitude?"

"Very nearly, yes," he replied, "For though we do not wish war, the principles at stake here are important enough for us to sacrifice an easy life for them. We've grown used to it, everything is done in such a way as to promote secrecy and stealth, those being our main advantages in the conflict. Out of hundreds of outposts like the one we were just in, for example, only four others have ever been discovered, and the Zards still have no clue where our fortress is." This he said in a boastful manner, but as he did a faint spirit of sorrow spread across his face for an instant, as if in memory of one of the raids of previous times.

"That explains their rapture when they found the false map," I returned, "But I must admit that I am still ignorant of the cause of the wars. It was said that it was conflicting ideologies, yet that is self-evident, as all conflict is at heart just that. I don't mean, either, the actions that caused the most recent inflammation, but what exactly your conflicting ideologies are? What is it that keeps you from harmony?"

"You have a knack for hard questions," he said with a smile. Then he paused for a moment to collect his thoughts. At length, he continued, "The Canitaurs have a profound respect for all that has gone before us, we honor the traditions of our ancestors and revere their beliefs and their ideas of truth. The past, in the guise of history, is the key to the future, we believe, and we hold strictly to the worship of Onan, the Lord of the Past," at this my attention was perked. He continued, "Our adherence to the ways of our ancestors is based on the idea that what has continued throughout the ages has continued because it is right, that it has remained steadfast because it is based on the immovable foundations of reality. We follow Onan because he is real, because the past has existed, and it is certain that it will continue to exist, and because that existence dictates the operation of the present. Although we may seem ritualistic and entrenched in tradition to the outside observer, we enjoy the comforts of knowing that we are on a well tread path, that we are not alone in time but in company with our forebears. We are called the Pastites because of our beliefs, because of our tradition based lives that instill in us a reliance on history, on the events of the past as a light by which to guide our own actions, as a road paved by the flesh and blood of our forefathers which leads to happiness and peace."

Bernibus paused for another moment, as if in contemplation once again, before he continued, saying, "The Zards are followers of the future, or Futurists as they are called. They believe that the past is just that, the past: the ignorant and selfish times of the unenlightened who were too shrouded by prejudices to understand the world clearly. Instead they place their faith in the scientific and philosophical ideas of the day, believing that while history and the past were delegated to the control of the unsophisticated whose ways were superstitious and outdated, the present contains truth in its pure form. Reform and revolution are their watchwords, for they tinker with the very foundations of society and life in an attempt to cultivate it. Zimri is their Lord, of the Future, and they follow him loosely, for he doesn't require the strict adhesion that Onan does, which suits their independent and relaxed world view very well."

He went on, in summary, "In a word, the Pastites believe that history, the reality of the past, governs the present and the future, while the Futurists believe that the future defines the present and the past."

"I begin to see the differences," I replied in a humble, questioning manner, "And yet they seem to me to be passive, secondary differences, the kind that result in a conflict of subtle disagreements here and there, argued over dessert like tariffs or taxes, not at all violent. How is it that they take such a prominent role in everyday life that they can only be resolved by force? What is it that takes it from the fireside to the battlefield?"

Here I was slightly taken aback by the expression on Bernibus' face, it was one of surprise mingled with apprehension and questioning. He said, "Then you do not know?"

"Know what?"

He laughed, "I take it you do not." Becoming solemn again, he continued, "Our land, Daem is on the edge of ruin, and has been for all of my life and those of many generations before me. About 530 years ago there was a great war on earth, one in which no restraint was used, no mutually assured destruction, for nuclear weapons came into the hands of those who cared not for any life, not even their own. Tensions were high for a decade, and in the following segregation, the peoples of the earth lost their personal connection with their enemies, and, as always happens, ceased to view them as equals, but instead as evil ones bent on their destruction. Things came to such a crisis that at last a little flame was lit and it grew and grew until it became a full scale nuclear war. The destruction was total: no one was exempt, as almost everything, and everyone, was destroyed. The only surviving place was this island, which is the sole habitat of the delcator beetle, a small insect that digests nuclear waste and neutralizes it. The first few decades were horrible, before the atmosphere recovered enough to return to normal, and in that time things mutated and grew gigantic. The trees and foliage, as you see, are an example of this, even the redwood trees of old were nothing compared to the trees of Daem. And the Zards and Canitaurs grew and changed as well, and, as we lived on either ends of the island, as we do now, our forms morphed into the separate forms that they now take.

"And that is where our conflict turned violent," he continued, "For it is our desire, on both sides, to return the earth to its previous state. The Pastites want to return through time and stop the destruction before it happens, because we believe that the past is what must be changed in order to change the present and future. It is the actions of the past that brought about the present woes, and it is they that must be undone. For their part, the

Futurists want to change the present through the future, to go into the future and bring back its completion, in the form of restored RNA cells, which is congruent with their belief that the past is the past and all that matters is that which is yet to come, that which still has the hope of existence."

I looked at him as he finished and said, "But, why not do both. Wouldn't that be more effective than fighting each other? How can continued destruction revert previous destruction inflicted in the same manner? Could not both ideas be tried?"

"If only they could," he replied. "It goes back to Onan and Zimri, you see, for we ourselves cannot do such things, but the gods whom we follow can. Shortly after the worldwide destruction, we, meaning both the Zards and the Canitaurs, received the prophesy of the kinsman redeemer, who would be sent to help us change the earth to its former majesty. He was to be one from the time right before the beginning of the final firefight, one of the ancients who still kept the pure human form. Our hostilities broke out in an attempt to control the entire island, so that when he should come, the dominant force would have him. Each side was convinced that theirs was the right way, the only way through which the end of restoring the earth's ecosystem could be reached. You are the kinsman redeemer, Jehu, for you fit the prophecy perfectly, and I am glad that you have fallen in with us."

After his discourse, Bernibus fell into a silent meditation, as did I, and the rest of our walk through the now dark wilderness was one of silence and solitude. Given the cessation of action in my narrative, I will take this opportunity to describe the circumstances of my arrival on the island of Daem, about which you are no doubt wondering.

Chapter 4
Onan, Lord of the Past

Not wishing to delve too far into my past or relate what would be mundane and disconnected with my story, I will summarize with brevity what my situation was. I was a military man, an Air force pilot to be exact, and was on active duty patrolling the no-fly zones off the coast of China, it being, at that time, an area of very high tensions. The situation was grim, as any small incident promised to set the pendulums of war into motion, but the worst had subsided, and things were beginning to look as if that incendiary incident wouldn't come after all. The main part of my story begins on a cloudy night of what was to me just a few weeks back, though it seems like many ages ago now, and indeed, it was.

I was flying over an area that was littered with small volcanic islands, the type that rise above or fall below sea level continually, so that what one year is above water is later below. Some of them have even been known to only rise above the waves for a short time, and then vanish from the sea completely, worn down by wind and waves. The night was murky, and the air was thick with water and dust, the result being that there was no natural light whatsoever, and any artificial light that could be mustered was largely reduced to nothing, visibility being no more than twenty feet.

The wind was calm and the flying, though strenuous from lack of sight, was without turbulence. I was doing well, until out of nowhere I heard a loud crack of thunder, followed by a bolt of lightning that hit the plane. At once I lost all of the instruments, excepting the actual control of the plane in manual, meaning that the radar and all the guidance systems were crippled, and I could see nothing. Not knowing what to do, and not being able to radio for help, I pulled down and slowed until I was just barely remaining airborne, and began looking for an island to land on.

Once below 200 feet, the clouds gave way and I saw an island. I aimed for it and slowed more, preparing to land on it. I did, though just barely, for it was extremely small, being one of those inconsistent volcanic islands. Getting out of the plane, I was greeted by a strong blast of wind that was dripping water from its cold grip, and I was instantly chilled to the bone. There was nothing on the island at all, except for the hole in its center, from which, no doubt, came the lava that had formed it. It was on a slightly elevated hill, and looked as if it had not erupted for many thousands of years. With nothing to do at that moment except to get an idea of the island that I had landed on, I walked over to it and knelt down beside it, peering blankly into its depths. It seemed to be absolutely devoid of light, and, as often happens, its darkness was mysterious to me, for I wondered what lay hidden in it, and my curiosity got the better of my common sense. I leaned slowly forward. Then, as I did so, I heard a loud and terrible voice, personified in the crashing of the waves and the moaning of the wind, and it said in a monotonous and unending refrain, "Enter." Nothing more nor less than the continual repetition of that word. This alarmed me, and as I did not want to do that, I began to stand upright and back away from it, to return to my plane. But as I raised my knee from the ground in order to stand, my other knee slipped under the increased pressure, and in the ensuing instability, I completely lost my balance and fell forward into the hole.

There are certain events in our lives that change the whole course of our existence, and falling forward into the hole was one for me. Its immediate effects weren't injurious to me at all, but it matured with time, like a good wine, and grew until it overcame me, starting the chain of events which would result in my demise. Yet not only mine, but that of everyone.

Let me continue, though, and I will explain what I mean and not confuse you more. I landed with a thud on a pile of soft dirt some twenty feet down, in a dark place which seemed open, not cavernous and cramped as I would have expected. My eyes adjusted to the darkness, and as they did, I realized it was not now totally lightless, for there was a faint glow coming from somewhere in the distance. Looking up through the passage I had come down, I saw that there was no way to climb up it, and, accordingly, set off to find the source of the faint light that came from the distance. After walking cautiously through the darkness, I reached a curve and then a tunnel-like exit to the spacious cavern that I was in, and as I turned it I saw the source of the light: lava flows. The room, or area, I had entered was rather thin and round, with a river of lava flowing downwards and a small ledge of rock winding along its edge. Together they descended spirally downwards at a gentle angle, taking the form of an intelligently designed ramp. As I followed it down I soon broke out in a sweat, for the gurgling, fiery plasma heated the area up to a warm degree.

I found myself looking intently at the flowing fire beside which I walked, its strangeness stealing my meditations from other things, and I looked at it absorbingly, not paying attention to the path that I walked on, so entranced was I with the feeling that its boiling character gave to me.

As I walked along the lava preoccupied with my meditations and not paying conscious attention to the path, my subconscious was carefully monitoring my way, and when once my eyes glanced upward, I quickly saw that my surroundings had changed. The narrow, spiral descending tunnel had given way to a very cavernous area where the lava flow formed a large lake of fire. A domed ceiling crowned this great room, though not exact and polished, having instead a rough appearance as it stretched from wall to wall, a semi-chasm of a hundred yards, more or less, with its uppermost height being not less than twenty yards. On the far walls were two lava falls, trickling from raised tunnels in the wall into the body of lava, which covered the whole bottom of the room. There was a platform that sat in the middle of the fiery lake, connected to the tunnel I had come from by a walkway of stone. This room was different than the other two, also, in its fashion, for while the previous had vague evidences of intelligent design, this one was very obviously artificially decorated. The walkway above mentioned was of ornate stone with an intricate design of circles, squares, and triangles carved into it, and on each corner of the center stage was a long pillar that reached from floor to ceiling, each carved like a totem pole, with a variety of animals and shapes stacked upon one another. The dome was done ornately as well, for I saw as I walked further into the room that what I had thought had been imperfections in the dome proved to be an elaborate three dimensional sculpture that stuck out from the ceiling, depicting an intricate scene of figures and telling a story of some great saga of war and peace, pride and prejudice, love and hate, faith and betrayal, all combined to make the greatest mural: history, the story of time itself.

As I looked in awe upon its beauty, I was startled by a voice coming from an unseen figure somewhere on the center platform. It said, "Jehu, you have come at last. Welcome."

The voice was very gentle and pleasing to the ears, slowly and confidently spoken, meticulously articulated. I looked around in its direction and saw a short, elderly gnome with a long white beard reaching to his chest and a short crop of hair on his oblong head, which was outfitted with a sharp, angular nose, a pair of sparkling eyes, and two protruding ears. He was no more than four feet tall, and no less than three, with a dignified poise to him, and was dressed in a dark robe with a black and gold design on it. We looked at each other for a moment, he smiling pleasantly and me expressionless, for though I felt that I should be surprised, or at least bewildered, at the sight of a gnome in an underground cavern, I was not, it was as if I had almost been expecting it to happen, as if in the back of my mind I had already been there and done that. Perhaps it was only a case of predestined deja vu, or maybe it was something less tangible. Either way, the gnome then broke the silence again, saying:

"Let me introduce myself, Jehu. I am Onan, the Lord of the Past, and these are the Chambers of History."

He then paused for a moment, waiting for my reaction, which was, again, not too much surprised, but rather complacent, thought I didn't look bored or snobbish, as is sometimes the case in that situation. Instead I became as genial as possible, realizing that whatever force was behind this, it was greater than I.

"Hello, Onan, it is pleasure to meet you," I said, advancing with a proffered hand extended towards him, which I realized belatedly made me appear oafish, but he took it good-naturedly, and with his pleasantness eliminated my unease at shaking the hand of one half my size. He then beckoned for me to follow him, and turned and walked to the center of the platform, where he unexpectedly laid down on his back, facing the muraled dome. I did the same, somewhat hesitantly, though I found it to be quite comfortable once I was down. He saw my sluggishness and by way of explanation said to me:

"Do not be troubled, my dear Jehu, for we lie on our backs to bring about clarity of mind."

Then he continued speaking, calling my attention to the sculptured dome:

"That is history," he said.

"What do you mean," I asked, "I've always viewed history as an organic being, constantly growing as it devours the present."

"It is an organic being," he replied, "A monstrous beast of sorts. But that (meaning the mural on the dome), my friend, is the genetics of history, its code that dictates what it is and what it will become, the master plan."

Allow me to take a moment to describe the mural for you. Firstly, its form: it was spread out across the dome like the painted ceiling of the Sistine Chapel, its whole being a broad, harmonious picture that complimented itself, telling a story throughout its united branches. It was much more than a painting, though, because it stood out from the dome like a group of completely independent sculptures, but placed so as to tell the combined story with a sort of native ease, not stressed or artificial, yet seeming as natural and beautiful as water in its flowing grace. Now I will endeavor to describe its content, though I realize that in this case the picture must be worth many millions of words.

The center of the mural was its beginning, and there a man was standing proudly upright, dressed in splendid clothes of fine linens. He held in his hand a magnificent cup of gold with a row each of diamonds, rubies, sapphires, and pearls running along its breadth. It contained a dark red liquid, which appeared to be boiling, and the man was holding it out to a fierce lion whose shoulders were four feet across and whose mouth was like a cavern, with stalactites and stalagmites of the most terrifying nature. With an evil glare in its eyes toward the man, the lion drank thirstily from the cup. Around the man and the lion there was a ring of blazing fire, leaping out of the dome like great pillars of flame, entrapping them within its narrow circle. On the outside of the fire was a group of mighty lizards and beasts, the smallest of which was larger than several elephants. Their whole attention was paid to a great fight in which they were engaged, yet their foe was naught but the reflections of themselves on the great sea which surrounded the island that held these strange sights. Several of them were dead or severely wounded at having been accidentally mauled by their fighting brethren. Across the ocean from the island there was another landmass, whose far edges were not in sight. On it were many ape-men bowing down in worship of a gigantic White Eagle which was soaring far above them with a multitude of lords and ladies gripped in its massive talons. The lords were dressed in silken robes and adorned with many pieces of fine jewelry, and the ladies were clothed in skirts of crimson; both groups had upon their faces looks of pleasure, and contempt towards those far below them.

Onan continued speaking, "You see, Jehu, the whole of history, both that now written and that yet to come, is planned, executed according to its own power, for the course of time is marked as clearly as the tides: by its own coming and going it is revealed. Revealed, however, only in an abstract and undefined manner, so that while its marks are clearly seen, it is only by special revelations that it is shown in a comprehensive and detailed light. And that is why I have summoned you here, my dear Jehu, for you are the chosen one, summoned to help me."

I was skeptical and asked him, "You summoned me? But how, I was to forced to crash land on the island by the weather, and accidentally fell into the volcano's mouth. It was by my own freewill decisions that the circumstances of my arrival here were fulfilled."

Onan laughed quietly and said, "History is not an unstoppable machine, allied with fate to control the destiny of all things past and future, nor does it nullify the power of man's freewill, yet the force that acts upon the minds of men to form them is history itself. You see, men are not the opponents of history and fate, for they do not impede its progress with their freewill decisions, instead they are its minions, its slaves, building up its strength and carrying out its dictates by its influence, so that they become history as they serve it, adding to its organism their own consciouses. While you were brought to these Chambers by circumstances of your own choosing, your desires in choosing those circumstances were dictated by the experiences of the past. But never mind how I summoned you, for you are here now."

"Very well," I said, not wishing to disagree with the Lord of the Past. Still, I was in a stubborn frame of mind, and asked, "But if the past is as powerful as you construe it to be, then why does the Lord of the Past need the help of a mere mortal like myself? Or do you mean you need a more direct agent than those you control only by influence?"

"Something like that," he answered. "You see, there was a great disaster once, which was blamed on me, and in order to atone for it, I promised to send a kinsman redeemer before anything so devastating happened again, and I believe you are the perfect choice."

"What devastating event hasn't been blamed on the past in one form or another?" I said, "But why not just go yourself?"

"It is against the rules," Onan told me.

"How typical."

"Yes, indeed, I sometimes wonder what good it is to be a god if you can't do anything yourself," he said with a sigh.

"What do you want me to do there, then?"

"I cannot tell you, unfortunately."

"Against the rules?" I asked.

"Very much so. All that I can do is send an agent with a slight understanding of the situation of history and physical existence to the people, but he must make the judgments of how to proceed all on his own. If I did tell you, it wouldn't be much different than going myself, and then there would be no human resolution to human problems."

"Our lives serve as a spectator sport to the gods, then?" I inquired of him.

"I am afraid not," he said, "It is much more serious than that. The Greeks were not all wrong, you know."

"Who else, I wonder."

"Not many," he sighed, "But tell me, are you ready?"

"As I'll ever be."

"Then I will begin. The understanding of life begins with the understanding of physical existence," Onan said, "And by physical existence I mean the quality of being materially animated. Not to confuse it with consciousness, which is the ability to think and reason, it is rather the realm in which one has substance and continuity. I will call the elements of physical being time and matter, those words representing widely known concepts. Matter provides the raw substance and time gives those lifeless objects a plane of being to exist in. Without time, matter can do nothing except sit in a sterile state, in a vacuum in which nothing could occur; and without matter, time would flow, but nothing would move with it. Thus, the basis of physical existence is time and matter, each being useless separately, yet together being the perfect combination of a tangible object and the fluid, forward movement to animate it. Imagine it as a three-dimensional painting, matter given depth by time."

"Not so complicated," I said cheerfully.

"Not yet, you mean," he laughed.

"Exactly, tell me more."

"Not just yet, Jehu. First you must help me."

"The time to begin has come then?" I asked.

"Yes, you must go now," he said, "And remember, I'll be watching. Good-bye."

And with that, not even standing up, Onan put me into a deep state of comatose and sent me through time to the unknown lands and people whom I was to deliver. I awoke, as you will remember, in the center of the savanna. Now that you know the circumstances of my arrival on Daem, I will go back to where I was before: on the way to the Canitaur's hidden fortress.

Chapter 5
The Treeway

I was walking in silence through the rugged forests of northern Daem alongside Bernibus the Canitaur, with his fellows Wagner and Taurus before and behind us, respectively, the former leading the way, the latter covering our tracks, and both on the lookout for an ambush. An entire lifetime of guerrilla warfare and privations of all kinds had instilled in the Canitaurs a strong and prevailing sense of caution, which sometimes rendered their lighthearted and almost spiritually frivolous nature to the casual observer a dense, deceiving demeanor used to conceal their true selves. But that was not the case, I believe, for they were, or at least Bernibus was, truly amorous in personality.

The sky was then in its deepest dark, and in the few breaks in the canopy above large enough to be seen through, there were few celestial lights to illuminate the depths of that mountainous forest. The forest itself sprawled like a great metropolis along the lands above the large central lake of Daem, Lake Umquam Renatusum, which was close beside the Canitaur outpost where we had narrowly escaped discovery and capture. However deficient in sight the forest was, it was abounding with sounds, everything from the call of the owl to groan of the bull frog, it was as if the whole of the forest had congregated about us, drawn to us by some unknown scent of interest and intrigue.

Continuing on for some time in the same way, I found myself growing weary, nodding my head slowly towards the oblivion of sleep, until I was brought to an instant liveliness by Wagner's announcement that we had reached our destination. I looked around carefully, yet I saw nothing at all to indicate the entrance to a large, covert military establishment, much to my companions delight. Their whimsical sense of humor surfaced once again as they laughed with seemingly infinite pleasure, both at my wondering expression and with a sense of satisfaction at their own cleverness. After the outburst had been subdued and a certain level of solemnity had been reached, Wagner approached the nearest tree and knocked on it with a rhythmic rut-tut-tut.

Expecting their old trick to be replayed, I waited for the tree to open, but to my surprise, it didn't, instead a strong rope ladder dropped down from a tree several yards to the east. This we climbed, and I found that I had been mistaken as to the height of the ancient wooden towers, for they proved to be even loftier in dimensions than I had imagined. Accordingly, it took us a good five minutes to reach its top at a quick and steady pace, and all through the climb I was terrified at the long drop, from which the ladder offered no protections. Yet I made it to the top safely, and found that there was a large platform built securely among its upper branches, with enough room to hold a few dozen persons, and there was even comfortable seating in the center. There were four guards stationed on the platform, each equipped with a long bow and a quiver of metal tipped arrows, and though they were hardly visible through the dim light emitted from the covered lantern that lit the platform, I could see them quietly conversing with Wagner and Taurus while Bernibus and myself reposed on the seats provided for that very purpose.

They conversed for awhile, though I could not hear them, nor could I see them well enough to judge their facial expressions, but Bernibus waylaid any anxious thoughts I had with his encouraging tone, and also by giving me a drought of ale and a loaf of bread to overcome my fatigue and hunger, both of which I quickly consumed. He gave me more bread, but wouldn't allow me another glass of ale, for safety's sake. At first I thought he deemed me easily overcome by spirits, but I soon discovered his reasons and thanked him.

Wagner returned from the guards and, finding that we were ready to proceed, led us to the far corner of the platform, where we were joined by Taurus. We then set off on a road that ran above the lower levels of the canopy, made from jointed platforms that were attached to the massive limbs of the trees, meeting the branches of the next tree half way across, forming a continuous, snaking path far above the ground. Traveling on those paths we made our way criss-crossingly to the west. The walking was no more difficult than on the ground, for the boards were firmly secured to the great branches, which were at least five or six feet wide, and there were short rails as well.

After no more than half an hour of travel on the 'Treeway', we reached another large platform in the center of a great tree which was very much like the first one, excepting that the trunk of the tree came up through its center and there was a door leading into the trunk. There were eight guards on this platform, but they let us pass without more than a friendly gesture, their scouts having, no doubt, seen us long before and ascertained our identity and intentions. They seemed to have been expecting the return of Wagner's group, though the addition of me they appeared to eye curiously.

Wagner led us directly to the door, which opened into a set of circular stairs that wound down the inside of the tree like the insides of an old world lighthouse tower. The stairs descended further than the tree ascended, wrapping around almost infinitely, at least to my wearied senses, which were depleted of vividness by the treacherous toils of the proceeding day. Down, down, down went the stairs, until at length we reached the bottom and found ourselves in a cave, the stairs ending in a small foyer area which opened out into the cave, it being delved into the bedrock layer, indicating that we had indeed passed below the surface on our descent. The passage was really a narrow defile with high walls on either side, impenetrable due to the fact that they were the foundations of the earth above. It stretched on for a ways, its whole length commanded by little, turret like stations which stuck out from the upper wall, in which were stationed groups of archers, and though they now stood in a solemn, dignified manner, any opposition that attempted to force a way through would have been decimated. Yet they stood at attention and made no noise or movement at our passing, instead being the essence of well disciplined soldiery.

This narrow chasm led onward for about three hundred yards, the walls stretching upwards in such a fashion that it brought to mind images of Moses crossing the Red Sea, with great walls of water suspended in air on either side, ready at any moment to come crashing down upon them, their lives in the hands of another. So did I then feel, the Canitaur guards being able to slay me on the slightest whim of fancy that struck their minds into a sadistic mood. Yet I was not afraid, instead I was overcome by a feeling of relaxation, where all cares and worries are given up as frivolous burdens, not necessary and not helpful, being, in fact, harmful to the mind.

The defile, or narrow passage, led to a great abyss, crossable only by a drawbridge controlled on the other side, which was at this time lowered and ready for us to cross, which we did, accompanied by four honor guards who were dressed in all the pomp and pleasantry known by the Canitaurs. It was a custom among them to greet newcomers with an honor guard which escorted them to the body of dignitaries and aristocrats that would be waiting to welcome them in style. This was done for us, and we were led into the fortress' great room, which was used for discussions and debates, via another winding stairway that took us even further below the surface. It was a splendid room, equipped with all kinds of luxuries and embellishments and spreading out like a quarter circle around a central stage with a podium upon it. Seats were arranged in arching rows, with a sort of cluster of seats around a wooden desk being allotted to each of the members of the council and his aide de camps; there were two hundred such clusters. Sitting there like they had been woken from sleep to attend to us were the delegates, looking tired and untidy, a rare state for a Canitaur to be in, with their clothes ruffled, their hair uncombed, and their eyes glazed with a discordant state of mind.

Wagner, who turned out to be a high official among them, led me to the top of the stage where the podium was, with a sofa, desk, and several chairs behind it, concealed from the council by the raised floor and walls that formed the base of the podium, creating a small, private anteroom for those at the podium. I laid myself down tiredly on the sofa to rest while Wagner took the stage and began to speak.

"Friends, comrades, associates," he said to the council, "I thank you for neglecting your beds at this late hour to join with us here in the Hall of Meeting, for there is something very important to be shared. You are all no doubt familiar with the ancient prophecy of the Externus Miraculum: long ago it was told that in our extreme need, when hope no longer exists in the hearts of many, an ancient would be sent by Onan our lord to redeem and deliver us from the evils of this world, for as our doom was wrought in their times, so would our hope originate. The past cannot be changed except by those who first made it, and our present is dictated by the happenings of the past, so that for a better future the past must be changed, and only then will we be freed from the burdens of history."

He continued, "We have therefore long awaited the arrival of our kinsman redeemer, who will change the past and prevent the cause of our current woes from happening, for without its roots, what evil can grow and flourish? Our redeemer was to come on the Kootch Patah, when our adversaries the Zards are not watchful, being drunk with celebrations at the turning of the year. Myself, Taurus and Bernibus went to the shores of Lake Umquam Renatusum, as is our custom, to watch for the coming of the promised one, and this time we were not disappointed, for he came to us, even as the prophecy says, as we sat hidden in the living tower. Seen by the Zards, we were almost discovered, until the promise of the hidden fortress drew them away, even as the prophecy says. And now we are here, delegates of the Canitaurian people, safely within our fortress with our kinsman redeemer, so what shall be done? Let us decide."

At this point he cast a glance towards me, as if desiring me to speak before the council, but I was in the last throes of wakefulness, where sleep has crept so far upon you that arrival in the land of dreams is only a matter of moments, and wakefulness is not desired, nor is anything else. I looked at him with my eyes glazed with that sweet, savory taste of sleep, and though I was conscious, I was not in control, only an audience to actions of my subconscious whims, and even that passed beyond my reach as my eyes fell shut, isolating me in the realm where worldly concerns mean nothing. And so I was when my exhaustion overtook me, leaving me sound asleep on the sofa behind the podium.

Chapter 6
The Fiery Lake

When I woke I was no longer in that room but in another, a small homely room where I was laid on a bed, the room being located, as I found out later, not too far from the Hall of Meeting. Though the depth of the fortress prevented me from knowing the time, it felt to be early afternoon by that strange internal clock that so seldom errs. It was correct, as usual. There was a quaint fireplace on the far wall of the room with a small, unadorned and unpretentious mantle, decorated like the rest of the fortress in a practical and experienced way, finding just the right flavor between the ornate, the practical, and the quaint, and avoiding all the while the clutter brought by superfluous material possessions. A table in the center of the room was furnished with a steaming meal, beside which sat my new friend Bernibus, smiling on me with a benevolent and almost paternal affection.

"Good morning, Jehu," he said, "Or should I say afternoon, for the morning has quite passed by already."

"Yes, and it has left in me a great appetite, my good man."

"As is shown clearly in your eyes," he jested, "Come and eat."

Needing no further urging, I leapt from my bed, sat down across from him at the table, and began partaking greedily of the hearty breakfast of hash browns and pancakes, which were pleasing to my mouth and stomach, for the tastes in food are controlled more by the condition of the body than by the time of day. When I had satisfied my needs, we reclined in our chairs and began conversing:

"Tell me," I said, "Did my untimely slumber yester eve cause any irritated prides?"

"Quite to the contrary, the council was well humored and followed your lead to their bed chambers."

"I am relieved to hear it, for I was anxious of appearing lax in ardor or animation."

"Not so, my friend, you are quite exonerated from doubtful thoughts. There is a session planned for this evening though, so may yet feel yourself put on trial."

"Unfortunate," said I, "But surely they can mean no harm, am I not the kinsman redeemer, after all?"

"Yes, you are," Bernibus said with a look of subdued apprehension, "We have an end in view, though the means are as yet not wholly decided. It is a complicated situation."

I smiled softly, "So is always the case."

"In truth it is: time reveals all things yet do all things reveal time?"

"What do you mean?" I asked him.

"Our situation is complicated by differing views of time, and I was wondering aloud if history and the present reality disclose the truth about time in the same way that time reveals the truth of the present. If our way were more illuminated, the journey would be easier."

"Perhaps that is why men look to the well lit paths of history, or to the dim conjectures of the future rather than the dark, yet detailed ways of present."

"Perhaps," he said, "But the present is so fleeting that it holds little intrigue."

"Even so, it is the stage, not still waiting behind the curtain, nor already performed."

"Yet the past controls by influences and prejudices, justified or not, and it will doubtless be the view of the council that the past must be redone, that the problems be addressed at the source," Bernibus replied.

"I am still in the dark about all your inferences," I said.

"My apologies, I forget myself. But let us not dwell on subjects which may become quite exhausted in the near future, for better or worse," he told me.

"Fair enough," I returned, acceding to the subject change, and jumping on the opportunity to steer it in a different direction, "I know little of you, Bernibus, so tell me all."

"There isn't much to tell," he coyly responded.

"Nonsense, Bernibus, tell me or I shall get very angry," I jested, imitating some mythological god's wrath.

He smiled discreetly and yielded to my request, "Very well, I will tell you. I was born in the year 490 D.V. (that is, Durante Vita), to a poor couple from the northernmost pier of Daem, the Gog."

"Wait a moment, Bernibus," I interrupted, "I didn't mean in that fashion, for when I say I know little of you, it is because I literally know little of 'you', not the circumstances that make up your past. I guess it goes back to the interpretation of the past and its powers, and since we can't seem to escape discussing it, lets embrace it willingly. You seem to believe that the events of your life have shaped you in such a profound way that their mere description is sufficient to explain your personality; I will grant that their influence has effected you subtly, but history is not the scapegoat of the present. The circumstances do more to define the character of an individual than to shape it, for even siblings with the exact same experiences can be greatly different in personality and achievements. But what I mean is this: your past has influenced your present, yet it is gone and your present remains, show me Bernibus, not his previous forms."

You, who are now reading this, may think this statement of mine to Bernibus to be hypocritical, in light of the very purpose and intent of these memoirs. You may be thinking that I am relating this whole happening in order to justify my actions and decisions. But that is not the case, for I understand that you have no power over me, I have long been dead in your present and your sentiments mean naught to me. In fact, I wish to tell of the circumstances I found myself in as much as of myself, so that you may have a retrospective clarity in visions of the future. You will understand that statement later on, but for now let me say that I wished to know the essence, the person, the consciousness of Bernibus, whereas I wish to impart to you my story, though ere its end you may come also to know me. I have no ambitions of material immortality.

Bernibus understood my meaning, and though he disagreed with its theoretical imputations, he humored me and did as I suggested. He pulled back his brow in a reflective demeanor, brought his eyes to mine and began:

"You desire me to tell you about myself without literally telling you of myself. I suppose you mean that we discourse on some variety of subjects, so that you can see who I am discreetly," he said.

"Exactly," I replied, "You say it better than I."

"Perhaps it is for the best, as you will draw your own conclusions rather than be given mine, and instead of my telling you what I would like to think I am, you would see what I am in truth. Strange, isn't it, that though we think we know ourselves, we very much do not, and it is only the unbiased observer who sees us as we are. You know, I was once thinking of writing my memoirs, and I would have, except that I was afraid that if I read them afterward I would be forced to see myself as I am and be horrified at the truth."

"Damn the truth," I said.

"You're starting to sound like a philosopher," he laughed

"And you a psychologist," I rejoined.

"And where would that place us on the scale of artificial intelligence," Bernibus jested.

"Following the footsteps of Jeroboam," I returned.

"Hmm?

"Oh, nothing. Tell me," I asked more solemnly, "What position does Wagner hold among the Canitaurs?"

"He is the Khedive Kibitzer, our ruler in that he leads the council."

"And you?"

"I am his brother-in-law, a relationship that our culture places great importance on, especially as he has no blood brothers. I become, in effect, his partner, though he doesn't accept me emotionally as one, only in etiquette."

"Why is that?" I inquired.

"Because, I am of weak heritage. His sister loved me, and I her, but to him there is no such thing as love, only business, the destruction of the Zards at any cost. No price is too high," he told me with almost a vengeful scowl on his usually pleasant features, it soon passed, though, and left no trace when it had.

"You sound bitter, Bernibus."

"My feelings betray me, yet I am not bitter, only disillusioned."

"You sympathize with the Zards, then?"

"Not at all, I do sympathize, however, with peaceful solutions," he said.

"Which is why Wagner disapproves of you, no doubt."

"Yes, mainly, but don't misunderstand me. I am not a closet Futurist, nor am I a strict pacifist, I just can't help feeling that there is another way. But I understand the selection of ideologies, how the stronger breaks the weaker to submission, and while one flourishes, the other diminishes, and I understand focus points, but I cannot justify their marriage."

"What you mean by focus points?" I asked.

"They are the culmination of conflict, where two sides meet and the battle takes place, not meaning necessarily an important or strategic military, civil, or commercial place, but one on which the fighting occurs, the result ending in the defeat or victory of the whole campaign. The focus point of the Zards and the Canitaurs exists both on the philosophical and martial levels. On the philosophical level, it is the question as to what is the proper solution for remedying our current catastrophic situation. On one side the Pastites wish to correct the root of the problem by stopping its realization in the past, the Futurists, however, would venture into the future and brings its stabilization and completion back. On the military level, our forces collide in the forests around Lake Umquam Renatusum, the northern mountains belonging to us and the southern plains to them. The lake itself is of little importance, yet whoever conquers it will conquer all."

"Interesting," I said, "But I do not understand how you seem to imply that I am your ancestor, while Onan seemed to mean the opposite, that you are my ancestors."

"It is strange and complex, and we understand very little of it, ourselves. The time for the council has come though, for our talk has dwindled away the afternoon. Perhaps some of your questions will there be answered. But come, let us go."

"Very well," I said, "Take me to your leaders."

From that room, the one I had awoken in, it wasn't very far to the council room. Exiting it, we turned down a short, closed hallway that opened into the concealed area behind the podium that I spoke of earlier. On the sofa where I had fallen asleep was seated Wagner and on a circle of smaller chairs around the edges of the area were seated about ten stately looking Canitaurs, clean and well dressed, according to their customs. They greeted me amorously, with a mixture of eagerness, excitement, and hope painted on their purloined countenances, taken from the sleepless spirits of several departed generations of war-hardened veterans.

Standing as we entered, they greeted me cordially, and, once the formal greeting of a short bow and a blessing was finished, we all sat down, they in their previous seats, I next to Wagner, and Bernibus in a small chair in the corner, away from the circle of the delegates. He, that is, Wagner, then opened our dialog:

"Welcome to the council, Jehu," he said.

"I was under the impression that the council was much larger," I replied candidly.

"It is, but this is the leadership; we felt that the clamors of a full legislature would be overwhelming to you at first. I know it still overwhelms me sometimes," he laughed, and the others with him. That explanation sufficed at the time, but I later found that Wagner had taken control of the council himself, and that it had no real power: it never met for more than ceremonial matters, the Khedive Kibitzer, Wagner, controlling the rest. But I get ahead of myself.

One of the others then interjected, "Our purpose now, Jehu, is not so much to make decisions as to inform you of the decisions we have already made, not that we mean to exclude you from our counsels, but we've been preparing for this moment, your arrival, for many years, since it was foretold long ago."

"Decisions with what end?" I asked of them.

"The reestablishing of an efficient and healthy climate, both naturally and philosophically, one in which tradition, history, and experience reign supreme," Wagner said in such a way that I couldn't help but think that it had served as an idiom of his for many years.

"A termination of the Zardovian conflict, then?"

"Essentially, but not wholly, as there are other, more complicated ends in view, less integrated with the format of a completely ideological conflict."

"Meaning?"

"Meaning that we wish to return to our original forms," Wagner said.

"Those being, I assume, the same as my own."

"Yes, you see after the Great War, the atmosphere was so filled with radioactive materials that all life was destroyed, except for that on Daem, which was protected because of our distant and isolated location, and the presence of a group of insects that neutralize radiation. They were overwhelmed in the first few decades, for though they were able to reduce the amount to make it habitable, we degenerated into what we are now, Zards and Canitaurs, based on our habitats, we being mountainous, forest dwelling folk, and they plains people. At first our ancestors grew to immense proportions, as did the vegetation on Daem, but we slowly returned to normal size as the radioactive material was consumed. I am surprised that Onan did not tell you about it all," he said, looking at me with a slight tinge of confusion creeping into his wayward eyes, formerly filled only with hope and excitement.

"I wish he would have," I responded, "But he said that it was against the rules."

"Ah, yes, I forgot about the rules there for a moment," he laughed, his countenance returning to its former gleeful appearance.

"A foolish law, no doubt, and from whom?" I said, availing of the apparent intra-personal deja vu, that is, the converging of the presents of our two minds into one idea, between Wagner and myself to cultivate a bit of sympathy in my difficult situation. But there would be no harvest, for Wagner checked his mirth and said:

"It was necessary, and the Council of the Gods did well to govern themselves more strictly."

"How so?"

"Well, during the Homeric period the gods really went at it, using humanity as players in their battles, like a game of chess, actually. Come to think of it, chess did originate in the realm of the gods after the laws. Things were quite a mess back then, though, with a whole horde of demi-gods walking the earth, and it ended up snuffing out the first flames of democracy and leaving monarchies for the longest time."

"Homer's stories were true, then?" I asked.

"Very much so, but after the laws of physical abstinence were adopted things mellowed out considerably, and men went back to their self-obsession, their material minds weren't yet weaned from the physical realm."

"So the very men who claimed mental superiority because they were free from superstitions and divine disillusionment were themselves victims of their own sophism, and while they thought themselves crowned with enlightenment, it was naught but the Phrygian caps of their prejudices toward the material state?" I asked, with more than the average dose of irony and feeling, both for my subjects and myself.

"Exactly, upon disinterested examination one finds the theater of human history to be one defined by a ludicrous melodramaticy, the soap opera of the gods," he answered. "But we digress far from our point, Jehu, which is a discussion concerning the implementation of our plans of action formed in preparation of our current situation."

"So I had surmised," I smiled at the reminder, "But tell me, what are your plans, and what is the current situation?"

"This is a time of fulfillment, with the events of many of our prophecies coming to pass. Now is a time of action and of hope. You, our kinsman redeemer, have come, and the time is ripe for victory and domination, ripe, in short, for a return to natural existence, harmony between forces interior and exterior. Our plan, my dear Jehu, is to attack the Zards swiftly and fiercely and break their strongholds like the walls of Jericho, literally."

"It sounds daring, certainly," I said, "But is it not overly so? I was under the impression that the Zards were much superior in force than the Canitaurs."

"In the southern regions, where you landed, yes, they are, but we rule the northern sphere of action. Our forces actually form a soft equilibrium that keeps fate's pendulum from straying from its neutral position, so that a military action previously would not have been predictable, with either side being capable of winning. Under such conditions war is avoided, but now you have arrived. The Zards, as well as ourselves, have been expecting a kinsman redeemer, you see, and our war has been kept from raging by the belief of each side that their god would propel them to victory with certainty by the sending of one such as yourself. Your arrival changes things, it marks the beginning of our dominance," he told me vaingloriously.

"The muted felicity I have witnessed about my arrival is explained, then," I ventured, "Excitement that the end is near and victory close at hand, yet that feeling subdued by the realization that a period of deeper darkness must first be gone through."

"Your words are true," Wagner replied, "And yet I have a great confidence in our plans, which have been matured through many years of careful deliberation. As the time will never be more ready than at the present, in the present we must act."

"What is your plan, then?" I asked.

"It is calculated to end in the conquering of the Zards, and as such, only an unexpected and unrelenting attack at the very heart of their strength will succeed. Anything less will only bring them to a full alert, and then any battle will have to be drawn out with excessive casualties on both sides. Therefore, we have decided upon an attack on Nunami, their capital city and main strength, being the center and majority of both their population and economy. Yet an outright siege of the city is impossible for those very reasons, it being so self-contained that it can resist bitterly, and its military is so clustered that it can be brought into action almost instantly.

"Considering those problems, it was deemed necessary to draw the Zards away from the city and destroy it in their absence, so that they are left destitute of the means of war and sustenance, and rendered weak. To do this, we have spent the last several years stockpiling huge quantities of liquid fervidus flamma, an extremely combustible substance. It is stored in an underground reservoir in the foothills of the mountains, connected via aqueduct to Lake Umquam Renatusum. When the time is ripe, we will empty it into the lake and set it aflame, and our calculations show the flames reaching a height of five miles for a length of six hours, which should be enough to gain the Zard's preponderance," Wagner explained.

"But wouldn't it catch the forest on fire and burn down your whole empire in the process?" I asked, alarmed at his apparent lack of vigilance.

"We have been treating the trees on a ten mile radius with an anti-flammatory solution for several years as well, and it is quite impossible to set them on fire."

"Which explains why you dared to have a fire pit in the trunk of a tree outpost."

"Yes," he laughed, "We aren't so foolhardy as we may seem. Appearances can be deceiving."

"The exodus of the Zards from Nunami is almost guaranteed by the mortal's natural curiosity and delight in the calamities of others," I said, "But how do you plan on leveling the town before the remnant raise the alarm and the mass of the people return?"

"Atomic anionizers," he returned.

"Which are what? They sound like they are beyond my level of understanding."

"Not at all," Wagner told me, "Do not be fooled by the technically complex sounding name. An atom is the smallest form into which matter can be broken down into while still retaining its identity, and an anion is a positively charged ion, or in other words, an instance of an atom in which there are more electrons than protons, resulting in a charge of negative electricity. An atomic anionizer is just what its name would imply: a device that morphs normal atoms into atoms with an extreme negative charge by emitting massive amounts, to the tune of many millions of moles, of solitary electrons into the air through a bombing device."

He went on, explaining the consequences of the weapon, "An atom, and therefore all matter, which is made up of atoms, is engaged in a constant revolution around the nucleus, in the same way in which our solar system revolves around our sun, and our sun around the black hole in the center of the galaxy. This revolving motion is the basis for the formation of all matter that we know of, both in its smallest form, like the atom, or its larger forms, like the galaxy. The electrons emitted from the atomic anionizer are drawn into an orbit around the nuclei of the atoms of all the matter near which they are detonated, much like the way planets catch satellites and space debris into revolving rings around them. This addition of electrons gives the atoms such a powerful negative charge that the poles of the atom, which regulate its rotations in much the same way that the earth's axis, or poles, regulate its rotations, are thrown from their natural equilibrium, causing the poles to reverse. This, in turn, changes the direction in which the atoms rotate, and in the brief instant in which the force of the revolving movement, or gravity, is not strong enough to retain the atom's shape, it lapses, bringing the materials they make up crashing down in disarray.

"We will plant some of these 'atomic bombs' inside the city of Nunami, and when they go off, the buildings themselves will implode and tumble to the ground. One hand-sized capsule can easily level almost ten square miles, and we have enough of them to bring the Zards to their knees, with plenty to spare for any circumstance."

"Wouldn't the bombs kill those who set them off, though?" I asked him anxiously.

"We have electron deflecting suits that negate the effects of the anionizers."

"I'm glad to hear it."

"And well you should be," he grinned, which, as out of place as it would seem, looked completely natural on his countenance, "For you and I shall be among the bombers. Our meeting must end here, though, my dear Jehu, for we each have things to attend to in preparation for the attack on Nunami. I will see you soon, until then, farewell."

"Farewell, Wagner," I replied, and we each stood and bowed as we prepared to depart, each to our own occupations.

With that our council ended, and, in the company of Bernibus, I was sent to another area of the fortress to be measured for an anti-electron suit, in order to protect me from the effects of reverse revolution. We didn't converse in the beginning of our walk, for my mind was too busy subconsciously thinking over what Wagner had said to have any conscious meditations.

We walked through the fortress towards the northern section, which held the technological rooms, so as to get an anti-electron suit in the making for myself. Realizing that the fortress has been little described, I will do so now. It was broken into six different sub-divisions, each branching from the only entrance, which was in the center of them all, the different divisions connecting to it through long, narrow defiles, or gorges, like the one at the entrance. This was for security, each area being independently contained within the whole. The six areas, or departments, as they were called, were as follows: the Northern was the technological and industrial research and production facilities; the Eastern was the residential department, containing also the civil services, such as medical care and distribution centers; the Southern was the agricultural and other food production areas, though there was little besides agricultural, for the Canitaurs were strict vegetarians; the Western was for mining minerals and other raw materials to be used by the other departments. The other two departments were below the others, being differentiated between by the names Left and Right, the Left being the governmental offices, and the Right the military headquarters, providing protections both civil and foreign (this was, incidentally, the beginning of the expression of the terms Left and Right to denote ideological preferences, but I digress). Uniform in all the fortress was the architecture, it being a strange mix between elegant and gentle arches and curves and brute practicality, for while the ceilings were high and open, and the walls wide, they were rendered homely by their plain surfaces and the absence of small triflings, conditions that were necessitated because of its identity:

an impregnable fortress containing a highly organized and self-sufficient governmental society, each citizen having a particular duty for the common good, and each kept from an unfarcical personal identity by the means of a statist society.

From the lower, governmental offices we went up a flight of stairs that wrapped round and round a tower-like tunnel, and soon reached the departmental portal. Once there, we took the northern tunnel, which opened into a large hall that stretched on almost endlessly, with hordes of tunnels branching off to the various agencies. There were a great many Canitaurs working busily, preparing for the attack on Nunami and its possible results, which, though long prepared for, had a few last moment components to be finished. Walking down the central through way, we went to the far end of the hall, which, as it was a walk of at least two miles, afforded plenty of time for observation and reflecting, two things that I am naturally given to. Accordingly, I turned to my companion, Bernibus, and offered in an almost philosophical way:

"Your society seems to be flourishing, though I am not surprised, as you all seem vigorously industrious. I am amazed, however, that no one shirks from their job, no matter how menial or trifling."

"We all have our assigned jobs, and all know that one slovenly job may cost us dearly," he said.

"I suppose I am prejudiced by my conceptions of personal liberty, but it is contrary to my conscience that the state should have more duty than to enforce the individual liberties by common force."

"But we are at war, and we must do as we do, or be trampled underfoot."

"If all states went no further than justice permits, namely the protection by common force the rights of individuality, liberty, and property, than there would be no room for conflict between states, and hence, no war."

"Yet it is our ideologies that bring war, besides, do not the ends justify the means?" he asked.

"Your ideologies may cause conflict, yet it seems that your behemoth states facilitate it into war. About the ends and the means, I don't know: I am no philosopher," I answered.

I sighed and was silent for a moment as we walked along, then, after a moment or so, I said quietly to myself, "I'm not much of a kinsman redeemer, either."

We continued on through the hall without further conversation, and I paid little attention to my surroundings, so that while my eyes saw and my mind displayed, my subconscious was not present in the effort, and thereby no memory was retained. This may seem to be the plot of an unimaginative writer to escape the use of that faculty, but as these are nothing but my written memories, and I make no claims of producing good fiction, I will leave that hall primarily to the minds of the reader.

Soon after, we arrived at our destination, which was very nearly at the end of the hall, and entered to find that we were expected and a space open for my fitting, which was soon accomplished, and my suit promised to be at my quarters the next morning. That would be just in time for the departure of the raiding party, which was set to cut out and embark for Nunami a little after that, in order to be in place in the hidden treetop posts surrounding the city before nighttime, as the operation was to begin at midnight. At first I thought that the attack was pushed forward in haste, but as I came to realize that my coming had been prophesied and a great amount of time had been spent preparing for this day, it seemed only natural that they should want to bring the hostilities to a close after such a long time. There were other considerations as well. The weather, for one, had to be dry and not at all windy for the fire to be safely attempted, and also the possibility of the Zards making the first offensive could not be ignored, for they had knowledge of my arrival and may have felt forced to act to prevent the very type of thing that we were about to attempt.

Chapter 7
Down to Nunami

When I awoke the next morning I found Bernibus and Wagner conversing quietly in the corner of my bed chambers, and as I first opened my eyes I saw Wagner looking at me with a blank, glazed expression, while Bernibus' was one of apprehension, apparently on my behalf. It seemed odd to me, but as Wagner became livid again quickly after his split-second lapse and gave me a hearty "Good morning", I thought nothing more of it. After his greeting, he continued:

"The day is ripe for victory, my friend, and the time is come for battle. We both have some preparations to complete, and so must separate, but we will meet again at noon in the entrance hall. Farewell until then," and with that he quit the room.

I looked at Bernibus, yet before either of us could speak, we heard a low, hollow grumbling, like the shaking of some building or foundation. He looked in my direction for a moment with an alarmed countenance, before I said defensively, "Tis but my stomach."

"Then we must get you some victuals," he laughed, "And I have just the thing to satisfy you and keep you so for a day or more: some mirus. It is our traditional energy food, for though its taste is bitter, its after-life is pleasant."

"And what is food except a servant to the body?" I said, "Let us eat."

"Very well," he replied.

And eat we did, for it was brought by a food service Canitaur on a tray, and I was surprised to see that it was a mixture of broccoli, spinach, and mushrooms, with a flavorless, glowing sauce. He was right, incidentally, for it was both bitter before and pleasant after its consumption.

"I know of the solids, but what is this sauce?" I asked of him.

"Carbon," he replied.

I looked at him and questioned, "Pure carbon? I have never heard of its having this use before."

"Your civilization was long ago and had not developed it yet."

"That has perplexed me, now that you mention it," I said, "Onan seemed to mean that I was going back in time to help my ancestors, but you say that I went forward, that I am one of the ancients."

He was wary for a moment, though if it was because of the apparent conflict, or because I was on a first name basis with his god I couldn't tell. He soon recovered his countenance and said, "It is a complicated question, and I believe you should ask Wagner the next time you see him, after the raid though, of course. The time of departure is nigh now, however, so you should put on your anti-electron suit," he said as he picked it up from the corner and brought it to me.

It was a subtle dark brown and looked more like a normal suit of clothes than an electron reflecting suit, but then again, I thought, why would it be a strange looking apparatus? Why would an advanced technological age necessarily be devoid of any sense of fashion, although that would be assuming that any civilization had ever had one. Fashion is more a characterization of a culture than a basic and unchanging principle, for a desert people would wear clothes that would be most uncomfortable to a people who lived in the snow. Clothes may not make the man, but the man certainly makes the clothes, and you can judge a person by what they wear so far as it is in their power to decide what that is.

After putting on the suit I found that it fit perfectly, and above that, I found it to be very comfortable, including the head piece, which formed closely around the skull and was not at all noticeable or obscuring. In fact, as it was made of a plasma that allowed everything through except lone particles, it was so uninhibiting that a moment after I had put mine on I had completely forgotten about it. The only other part of the suit that stood out at all was the long, metallic buckle that secured the belt, it having a bowie knife hidden within it in an unnoticeable and inconspicuous manner. Bernibus had put on his as I had put on mine, and as I looked away from the mirror that was opposite the door, I saw him dressed the same as myself, yet because the suit so blended with his fur, it was hard to tell which ended where.

Finding that we were both ready, we repaired to the entrance hall. Along the way I asked Bernibus of his wife, Wagner's sister, of whom I had heard little and seen nothing. He was quiet for a pause, and then said:

"She was an angel, what else can be said?"

"Was?" I asked hesitantly.

"Yes, she was killed by the Zards on a border raid, as we were at that time living apart from the Canitaur mass with a few friends. She was less aggressive than her brother, and, much to his disapprobation, we lived with a group of separatists, believing that war, physical conflict, is never the right answer to ideological conflict. Wagner excommunicated us in his anger, though his sister was very dear to him, and after she died he was struck with remorse and made me his deputy Kibitzer. He felt that it would somehow do her honor, as it would recognize us as having been married and make me his brother-in-law, which is an important relationship traditionally, as he has no other siblings. So here I am, technically second-in-command, but because of my soft lining, I have no real command."

"You would not attack Nunami, then?" I asked.

He chose his words carefully, saying, "More pain will not negate the pain already in existence, yet war is not always avoidable, and sometimes it is even necessary."

When we reached the entrance hall, where the raiding party was to meet, we found that there was already assembled a majority of the force, including Wagner. The party was only twenty strong, as the atomic anionizers were to do the main work and the planned raid required stealth and secrecy, not force or might. Within a quarter of an hour all the stragglers had arrived and all the anionizers were accounted for, so Wagner gave a short debriefing to ensure that all the members were on the same page. We were to sneak into the city when the populous was distracted by the fire on Lake Umquam Renatusum, which was to be started at midnight. We would plant the atomic anionizers at the right spacing so as to bring down the whole city once we were escaped, using the remote control provided for that very purpose. The suits would protect us from the blasts, and, as a precaution, the remote had an automatic five second delay between being pressed and exploding the bombs, though it was more for form than practicality. After he finished we set off, being arranged two abreast per row, there being ten rows. Bernibus and myself were partners, for we had become close friends in the few days that I had spent among the Canitaurs, while Wagner was once again the leading guide and Taurus the rearguard.

After crossing the chasm that separated the hall and the entrance tunnel, we came to the long defile that formed the latter and passed through it swiftly, the lofty archer guards remaining as stern and immovable as when I had first come through. We then came to the winding stairs that occupied the hollowed innards of a massive and ancient tree, of which kind many were to be found in Daem, being at least fifty feet thick and 700 feet high, such gigantic trees that were never seen elsewhere, yet constituted the whole forests of the northern lands. I found that the stairs were as long as I had remembered, taking us a great while to ascend to the top of the tree, and when we had made it, we, especially myself, were dazzled by the effulgent light of midday. After having been out of the sun's reach for the last few days I was completely unprepared, though the shock helped me by curing me of the disillusionment that comes from not seeing sun, moon, or stars for any length of time. Taking a rest for a few moments on the seats on the platform, we collected our strength. After our brief repose was completed, we set off again with renewed vigor across the treeway on which I had first come to the Canitaur's fortress. You will remember that the road was made by the securing of five or six foot platforms to the intertwined branches of those great trees, over which one could travel with ease and be safe from exposure to those below by the thick foliage that grew on the trees and was carefully manicured for that very purpose.

Soon we reached the first platform I had seen, which we had come upon from below, but we did not descend there, instead keeping on by the treeway in the direction from which we had come that night, that being southward, towards the lake, the savanna, and the Zardovian capital, Nunami. The air was warm, with a slight breeze as we went along, and that, mixed with the plentiful flora about us and the songs of the treetop dwellers, rendered the whole feeling of the walk peaceful and happy, though its end was not to be such. I soon forgot the worldly concerns that plagued me as I was soaking in the simplicity of nature, not a simplicity of form, for all things are incomprehensively complex, but simplicity of meaning.

After a time I began noticing changes in our surroundings that indicated we were drawing nearer to our goal, namely, the trees lessening in proportions, the terrain becoming flatter, and the air growing moister and more vibrant. Still, the trees continued to spring up from the ground like great earthen tentacles, for while their size diminished, it was not by enough to change their demeanor, the trees anywhere on Daem being great in size.

The sun journeyed with us, and by the time we reached Lake Umquam Renatusum, twilight's last agony was being performed in the heavenly theater, and the rippling waters mirrored it, adding only a strange, flowing texture. The lake's current caught my eye with its subtle oddity, for it was amiss and it appeared upon close inspection that there was an undertow, as if there was an underground river flowing into the lake and bringing about its swirling currents.

Bernibus saw me looking down at the waters from the lofty road with a puzzled look, and asked me if I was wondering about the water's current. I replied that I was, and he told me that it was the fervidus flamma being pumped into the lake through the underground aqueducts, which, of course, was for the purpose of igniting it to decoy for our raid. Once it was explained it made sense, yet I looked at it anyway, for it was still a gorgeous and inspiring view.

We were moving quickly, however, and it soon was out of sight, and I again turned towards our destination with apprehensions of failure. They seemed to place great faith in my presence, as the emissary of Onan, and while I was, I was also Jehu, and I wasn't confident with my own abilities. But it was upon those the situation mostly rested, it being the resolve of the gods after the Homeric period to take a more removed role in the lives of men. I wonder how many from my own times were divine agents, for better or worse. Either way, my main concern then was making the correct decisions, for I rightly believed that my involvement would decide the matter, although not in the manner I had anticipated. As I looked about myself to reconnoiter the feelings of my comrades I was fruitless, for they all wore impermeable countenances, though that was itself an indicator of their resolve.

Within an hour after the fall of darkness we reached the outskirts of Nunami, or rather, its edge, for it was walled in with massive stone walls and battlements, with a sturdy gate of twenty foot width being placed at the northern, southern, eastern, and western ends. The trees hung right over the walls, and as such we were able to take positions from which we could descend into the city when the time to do so came. Yet we were still rendered invisible by the thick foliage.

Night's zenith blew in slowly on the wind like the belabored breaths of a dying man, and after a period of worry, it came: midnight, the appointed hour. No sooner had the moon reached its utmost height, shrouding the lands in a shadowless vortex, than a great blaze erupted from the northern lands, and it rose almost instantly to its estimated height of five miles. It was a terrible sight to behold, for any flame is a captivating display of inorganic life, but a pillar of flame several miles high is more than just an enlarged specimen, for it plays host to a great horde of phantasmal apparitions that wrestle ferociously with one another. As the flame shot upwards it cast a great light down on everything that rivaled the illumination of midday. At first I feared lest the light should show our silhouettes to the Zards, as we were between them and it, but it did not, or at least they took no notice of it if it did, for we were quite undetected in our hiding place.

Our worries were far from over though, for now came the crucial point in our plans: in order for our small force to infiltrate the city and place the atomic anionizers, the Zards must not only have been distracted and preoccupied with the blaze, but they had also to leave the city almost empty and go to the lake itself, for if a cry was raised, or any substantial resistance attempted, the complex procedures to detonate the anionizers properly, so as to level the city but not the surrounding country, may have been hindered. There were several factors on our side though, the element of surprise being the foremost, for in their excitement the Zardovian resistance would likely mistake us for a regular sized army and flee in fear at our supposed superiority, especially since the presence of me, the kinsman redeemer, was known to the Zards. Also, the Zards were known to be curious and careless and ruled by the desire for excitement, meaning that if an entertaining undertaking was possible, they would pursue it, no matter how dangerous or ill-advised.

Within a moment after the flame was lit, all of the Zards outside, which were many, were gazing with silent wonder at it, and in the second moment, all the rest had joined them in their confused contemplation. But the third moment witnessed a drastic change in their behavior, for their initial bewilderment wore off and suddenly, with a united prelude of the drawing in of a breath, they all began speaking at once, resulting in a clamorous din that lasted for a few moments, before things hushed again and we could hear a few individual voices discussing loudly. Though we couldn't make out their exact words, they were apparently conferring with one another about what action to take. Our breathing became slow and heavy and our brows were knit tensely, for we knew that the fate of our mission rested on what they did then, whether or not the long planned decoy would work.

It was an anxious moment, and one with a heavy burden attached to it. Fortunately, though, as our fate was decided, it was done so in our favor, for the Zards began exiting the city in a great multitude of scales that swept along the savanna like a tidal wave over a sandy coast. They came out fast and strong, and through each of the four gates, though only the northern was fully visible to us, the others being too far to be seen distinctly. Still, we could see them rushing out of Nunami at a quick pace, not hurried, as if frightened or finicky, nor slow as in deliberation and meditation, instead it was a steady trot that they took, allowing them to move safely and swiftly.

The tide of Zards swept steadily past us, and it was a good half an hour later that the final ones had left the gates and the city far behind. Most had taken some type of weapon, a pitchfork or club or occasionally a sword, for the threat of war was a constant, but none of them had any idea that their only danger was behind them. It was not all in the clear though, for a patrol of guards equipped with long spears and clothed with a tough, leathery armor were making their way to and fro along the tops of the walls, where there was a platform of about five feet across that served as a road to the soldiers in their watches. It was evident by their countenances, though, that the guards now on duty were more interested in the fire than in their immediate vicinity, thinking, no doubt, that the laurels were to be won there and not at Nunami, and as such, they paid little heed to the walls, instead walking with their necks craned precariously to the north.

We were able to jump unto the wall silently from our concealed roost on the treeway when the nearest patrol had passed by. From there we went along the wall a short way until we came to a battlement, there taking the downward leading steps that brought us to the ground. Once there we were pleased and hopeful at what we saw: everything was abandoned, and no Zards were in sight save those on the walls, whose gaze was cast elsewhere. We set to work, then, according to our preset plan, which was to break up into groups of two and cover the city with our atomic anionizers, so as to spread the destruction as evenly as possible. Wagner and myself were partners, and we took the central district, near the government's center, the palace, and the Temple of Time, which rose above the city like a great tree amidst a desert. It was, in fact, the very structure that had so stood out to me during my journey through the prairie upon my arrival, and once again its sobering sensation struck me, and I found myself staring up at its top, a full 800 feet high, the bottom being an ornate and elaborate temple. The middle, which supplied most of its height, was a long, round tower, and at top there was a spherical pinnacle which had what looked to be a room in it.

Wagner soon called my attention back to our work, and we busied ourselves with planting a bomb at the base of the palace, using a smaller type anionizer, which, I noticed, was set just right so that while all of Nunami would be leveled, the temple with its great tower would be beyond the impact and left standing. Just as we had set it correctly, we heard a high-pitched whistle, which was the preconcerted signal among the raiders to use if any danger was nigh. We looked up directly and saw its reason: a squadron of Zards had been garrisoned inside the palace and had not left like the others, apparently because its sole purpose was to protect their king, who did not leave the city, being preoccupied with business and not seeing the flames. When he did go to the window, he saw the fire, and rushed to see what was about, but instead of finding out, he ran into us, who were right outside the palace.

Wagner dashed wildly through the streets in an impressive show of dexterity, and did a wall-jump between two lofty buildings to gain the wall. The others had done likewise, having been trained by a lifetime of conflict to have nerves of lightning speed and earthly strength. Their instincts had come in subconsciously when they had seen the cause of the alarm and they escaped, without thinking of me in the critical moment. I lacked such strength and speed of mind and was caught as soon as I had seen the squadron, aided, probably, by the fact that upon seeing me the king had become excited and rushed at me with great speed. When Wagner had first turned around and saw me their prisoner, he looked crestfallen and hopeless, for he had no way to rescue me. He held the remote control for the atomic anionizers in his hand and was about to set them off and make good the plan, but before he could, our eyes met for an instant, and we connected beyond time and space, experiencing a strange intra-personal deja vu. All was silent and still in that instant, and I saw him struggling inwardly: would he detonate the anionizers and make good his long awaited plan, or would he retreat and leave the city unharmed, for though I was wearing the electron reflecting suit, the collapse of all the high rise buildings would litter the ground with debris from them, and all on the ground would be crushed. Would he spare me from death, or his people? In that instant his face spoke more than many others' do in their entire lifetime. It was cut through with a contrasting countenance, and yet inside of his eyes there was something foreign to them shining through, something that I had never seen on his fretless features before: evil intent. I could not tell if it was natural to them and simply well hidden, or if it was an alien expression, but it was fearfully expressed, and his eyes seemed to say, even at that great distance, that he took a third course, that he would save me, but not for my sake, instead for his peoples'. And then it passed, for he looked away, replaced the remote to his belt, and leapt to the ground, where the other Canitaurs were awaiting him. I saw him no more until the situation was much changed.

Chapter 8
The Temple of Time

I turned slowly away from where Wagner had disappeared over the side of the wall and faced my captors, the Zards. Chief among them was the King, he being a foot or two taller than the others, with a graceful and powerful pose that struck awe into the eyes of the beholder with its innate command and dignity, both of which flowed from it as naturally as water from a well. There were about twenty guards in the squadron that protected the King, but it was not so much from the terror of them that the Canitaurs fled, nor was it because of the guards that patrolled the walls and were sure to join any fray attempted, it was instead an apparent fear of the King, and rightly so, for his demeanor was fierce and sophisticated, as if he were not just a warrior nor solely a scholar, but a mixture of the two that gave him an aura that inspired fear, some unseen presence that filled the air around him and sent his neighbors into a reverencing awe reminiscent of a lover's sacred euphoria, intangible yet undeniable.

As I turned to him, he smiled and greeted me softly and pleasantly, in such a way that seemed contrary to his nature. Instead of being terrible and glorious like the crash of thunder or the din of waves, his voice was melodious, subtly so, like a soft summer rain affecting the dreams of a slumbering child as it falls gently on his face. There was a rhythm that ran through it, like poetry, yet not like average poetry, where the rhythm is forced and the lines deformed to its ungainly warble, but like heavenly poetry, where the rhythm is beyond the conscious and into the subconscious, where it inspires a feeling of quaint remembrance of itself, as if it were there and not there at the same time. And while it was soft and pleasant, it was not feminine, for it was a strong baritone, reinforced by its own superiority and strengthened by its wit and sobriety.

"Greetings, o' chosen one," he said to me, "I see that you have arrived safely."

"Yes, quite soundly," I replied, a little taken aback on two fronts: firstly that he was not angry or indignant that I had attempted to destroy his kingdom and take his life in the process, and secondly that he seemed to expect me, as if I were his midday tea partner.

"I am glad, for I would wish you no harm, though your Canitaurian friends obviously felt no such concern. But just as well, for they always were unpredictable. I'm sorry that there is no one here at the moment, or we should have a great welcoming parade for our newly arrived kinsman redeemer, but they are off at the lake, inspecting the fire I suppose. I must admit it caught me off guard for a moment or two, and at first I was actually quite surprised. I soon remembered, though, that our friends the Canitaurs would have gotten some notions in their heads of a battle, at your arrival. It must be a grand sight in any case, and not one to miss."

I gave him a strange look, for I was a bit confused myself at the attitude he donned towards me, very friendly, as was Wagner, as I recalled, though it seemed as contrary to his nature as it did to the King's. He saw the expression of my eyes, and seemed to read right through my thoughts and see my apprehension of punishment, for he beckoned to his guards to leave us alone. They moved quickly and uniformly, a well-trained unit, and positioned themselves in a line formation along the street. The King and I then strolled down their midst, they walking along with us at a distance of a few yards, which was all that the closely built buildings would permit. In a moment or two we reached the Temple of Time, which was on the far side of a large square plaza that opened up between it, the palace, and the government center. Once we reached it, he led me inside and the guards took up post around its outside.

"You need not fear," he told me when we were alone, "You are among friends here. You see, the Canitaurs were not the only ones waiting for a kinsman redeemer, the Zards were as well. That day that you were seen going into the Canitaur's outpost was a big disappointment for us, I had almost begun to think that you were beyond our reach. I am sure you know all about the conflict between us, and the circumstances of your time that brought its beginning about?"

"Yes, I do," I responded as we walked through the great entry hall of the temple, lined with bookshelves and a rich red carpeting. He was silent for another moment as we crossed into another room that led to a chamber with a long table in its center and a great many statues and works of art scattered throughout its whole. There was an altar at the far end, built into a giant statue of a White Eagle that graced the entire wall, it holding the altar in its giant claws.

He saw me look at it and told me, "This is the Hall of Time, and that is the altar to Temis, the God of Time. It is a very sacred place, to both us and the Canitaurs, for it was built by Temis himself, before the race of man inhabited the earth. By the time any men came to live on Daem, it had been buried by the dirt and debris of thousands of years, but when the Great

War took place, the shock uncovered it and revealed it to men, a sort of revelation that came only as it was needed the most. Daem's war started over the control of it, and to a point still is. To a certain extent is has helped us greatly, since the Canitaurs are afraid to lay siege to us in the regular fashion, for fear that it will be laid to ruin, and then our fate sealed in flesh and bone as well as earth and stone. But come, there is something I want to show you," he told me.

With that he started over to a door in the wall adjacent to the entrance, which, as there were only two doors, was the only other exit. It led to a long, winding stair that went up to the top of the tower that I had seen from below. We walked up it in silence, more from awe of its magnificent construction on my part than fatigue in climbing its steep stairs, which wound on and on almost indefinitely. There were no windows in the tower, and only a few paintings to liven up the sparsely decorated walls, yet they needed no adornments, for they were beautifully constructed from a strange stone that split and colored in a marvelous twisting pattern.

At last we came to the top. It was much like it had appeared to be from below, for it was a large glass sphere that sat on the tower, like the dome on top of a light pole. It was divided in two, and the stairs went right through the bottom half and opened into a circular foyer that then had a small flight of stairs running up to the main room. There were little closets and such in the empty spaces on the bottom floor. The upper room was a good thirty feet in diameter, and the walls and ceiling were all made of glass, very sturdy and insulating, yet completely transparent. On the floor was an odd carpet that was smooth and thin, like a silk or fine linen, yet very strong. There was a rounded table on the side of the entrance hole opposite the stairs, and a curved couch that sat against the wall behind it, cut perfectly to its circular outline. Two cushioned chairs sat at the table and a small end table leaned up against the couch, on top of which there was a medium sized spyglass, that is, a telescope.

The sun was just coming up and shining its golden hues on the surrounding lands, which were beginning to darken as the fires of Lake Umquam Renatusum died down to a faint glow in the center of the forests of the near-north. It was the first time that I had gotten a bird's eye view of Daem, and I was amazed at its beauty. The plains stretched on one side of Nunami like a broad field of gold in the morning light, its dew drizzled grasses waving in a solemn and dignified manner to and fro like the constant beating of the earth's heart, and when looked upon abstractly it moved as if one great beast of benevolence, holding itself in unison as it chorused back the silent tones of life. Its edges draped down to the ocean like a curtain of woven sunlight on the eastern and southern sides of the island of Daem, and on the western side of Nunami the great forest came up right to its edge. There was a little of the forest between the ocean and the city on that side, while to the north there was a great stretch of trees, all the way until the ocean again came into sight in the far, far north. On the ground the trees of Daem seemed like mighty towers and battlements of nature, and on the treeway one felt suspended in air hundreds of feet above the ground on a cloud of green and growing foliage, but from afar and above they were revealed in their true splendor, shooting up from the earth as if they were the arms of the ground itself, grasping huge clusters of leaves and branches far above in their tightened fists. Some way into the forest, the ground sprang up into mountains that were as fierce and behemoth as the trees that clothed them. They were terrible to the eye and mind, as evidences of the power that exists outside of oneself.

The city of Nunami was also revealed to me for the first time in depth. As I have said, it was surrounded by a thick, tall wall made of stones and precious jewels, with four gates, one at the furthest extreme in each direction. It was a circular city, made mostly of the same materials as the wall and temple, which were a plain, silvery stone; a dark rock with inherent patterns; a mixture of cobblestone and a colorful compositor rock; and a vast array of metals, everything from brass to silver to platinum. Made in an ancient style, the buildings were tall, the average being what was equivalent to at least a dozen or two stories in the pre-desolation times, and they were close together, built along roads paved with cobblestone and lined with trees whose girth, though not as monstrous as those in the wild, was still great. There were farm fields and vineyards and orchards and meadows for grazing animals all within the city walls, and not just congregated around the outside, for there were buildings all around the wall's perimeter, but scattered among the other buildings in a natural and pleasing way. In the southern part there was a lake that was of fair size, and a fleet of fishing boats anchored at its shore showed that it did its part to contribute to the city's well-being. Several of the trees throughout the city were especially conspicuous in their grandeur, for they rose hundreds of feet from the ground and had great waterfalls flowing down from their tops, as if they were crying great torrents of tears down from their aged faces, though if in sadness or joy, I couldn't tell.

To the east there was land visible from the height at which I found myself, though in the distance it became hazy and I could not make out its distinct features. It was evidentially corrupted, however, for it had an uneasy look about it, as did the ocean, which was a faint, pale shadow of the rich blue it was in my childhood days. The sky as well was tainted, and it looked to be filled with the accumulated atrocities of countless generations.

The clouds were thick and bluish, and the spherical mural of the sky itself had been greatly dried, cracked, and crumbled since my time, for it bore the marks of pain, the marks of the labor pains of the earth's last gestating doom. And well they should, I thought, for in the years since my natural life it had seen much suffering and much destruction.

The King broke the silence, saying, "Lovely, isn't it, Jehu? And it is all yours for the taking."

"What do you mean," I asked him.

"Exactly what I said, the whole world is yours, if you want it."

"But how?"

"All you have to do is join us, the Futurists, and we will reward you with all the power and glory that you can imagine."

At that I sobered up and replied, "But what of Onan, of my quest to stop the doom of humanity from materializing in this final juncture. He is the one who sent me, and he is the Lord of the Past, whom the Canitaurs follow. I am his agent, why would I turn from him to serve mere mortals?"

He laughed a slight, sarcastic laugh, "Tell me, Jehu, to whom did he send you, your ancestors or your offspring?"

"To my ancestors," I said slowly, "Though the Canitaurs seemed to imply that my time was long ago. To be candid, I do not understand."

"Of course you do not understand, and how could you, when no one has told you? You see, Jehu, the question of time is not so linear as you would think. You know full well that the conflict between the Zards and Canitaurs is over how to address the renewing of the earth: they would send you, our kinsman redeemer, back into time to prevent the nuclear wars, while we would send you to the future to bring back its completion. They hold to traditions as if they were the foundation of life, while our people have no traditions in the traditional sense, if I may use that oxymoronic phrase, but we look to what will come instead of what has passed. History is unimportant to the present, Jehu, because we have advanced to the point that we do not make the same mistakes as our ancestors. In the past, they waged war needlessly and did so in the name of humanitarian deeds. But today, we are advanced enough that we use peaceful and just means to reach our ends. In your day there were many absurd beliefs, for example the so-called 'fats' that were so vehemently avoided, are actually quite healthy, while on the other hand, protectionism and socialism are quite absurd ideas, and yet they were held dear. But today we have no such presuppositions, today we understand the world and know justice where your society knew only its shadows. We do not need to be bound by the mistakes of yesterday, for we have the enlightenment of today, and while the Canitaurs cling to the old time's ways, we have progressed to the point where we have no need of such traditions."

He continued, "It may seem to you foolish to follow Zimri instead of Onan, because Onan's realm has already been established and grows greater everyday, while Zimri's doesn't exist and never will, but you miss a very important point in the understanding of these matters. For, as you probably know, time and matter are the foundations of physical existence, and while the two components are independent, they are also parallel. Matter is always revolving, from its simplest form in the atom to its greatest in the universe, everything is revolving and rotating. So is time. Imagine time as a galaxy, revolving continually around the black hole at its center, that is, an enigma that is actually devoid of all matter. Time is revolving around a great enigma as well, which is devoid of time, that enigma being eternity. Eternity is not a place where there is infinite time, but rather a place where there is simply no time, it is the counter-part in the temporal realm of a black hole in the material realm. And just as a galaxy in the material realm revolves around the black hole at its center, in the temporal realm, the flow of time itself revolves around eternity. That means that time repeats itself over and over again, just as on earth a year is the amount of time it takes the earth to revolve around the sun once, in the temporal realm, an age is the amount of time that it takes the time continuum to revolve once around eternity. Just as every year the climate on the earth is similar, every particular day having its usual temperature and weather, and every general period having the same seasons, so is time. While every age is completely new and original, they all follow the same pattern, and through every age the same general events happen, though a few of the small details change from one time to the next.

"So you see, it is true that Onan sent you to both the past and the future of your original time. The Pastites would say that you were sent forward in time, because you existed in our past, while the Futurists would say that you were sent backwards in time because you existed in our future. While this would seem an unimportant question, it is not, for we have to choose one or the other. You, the kinsman redeemer have to choose one or the other. That is why you were sent, you have to decide. Our fate must be decided by a mortal because the gods have vowed to never interfere directly in our ways again. You must decide, Jehu, for you hold the fate of humanity in your hands: in all the other ages before us, the wrong decision was made, and every time some great calamity came that somehow threw the earth into a great ice age that destroyed all life for many millenniums. We know that the wrong decision was made, but we cannot tell what it was that was done. Tell me Jehu, will you join the Futurists? Surely you can see that the Pastites are just that, stuck in the past, with their obsession with traditions and legends. They are of the past, but we are of the future, we are the progressive ones. Dear Jehu, choose the future, and when the earth is spared from the great impending doom, we will set you up as ruler of the world to show our gratitude. Will you join us, friend?" he asked me with the most entreating eyes, though of somewhat doubtful sincerity.

There was a deathly silence that followed, for I was thinking long and hard about what I should do, until at last I spoke, "Your majesty, I am afraid that I will have to turn you down and remain with the Pastites. Onan sent me, and it is Onan whom I shall follow."

The King shook his head and sighed dejectedly, for a moment he looked disheartened and crestfallen, but then he again resumed his former prideful pose and said to me, less humbly and entreating than before, "Very well, I was afraid that you would do that. I have no choice now but to keep you here indefinitely as a prisoner, until such time as you realize the error of your ways and repent. It may seem improper to refuse the decision of the kinsman redeemer, but I must, for I will not allow my people to be destroyed by your ignorance."

With that he turned and walked quickly down the stairs to the door, turning to me just as he reached it and adding with an almost spiteful intonation, "But then again, what clarity of mind can be expected from someone from the unenlightened past." He then left the room, closing the door with a powerful thud, after which I heard a small metallic click and his strong, commanding footsteps fading down the long stairway. As soon as the sound had died away and he was no more to be heard, I ran down to the door and tried to open it, but to no avail, for it was locked. There was no way to escape: I was a prisoner of the Zards.

Chapter 9
Mutually Assured Deception

The light of the newborn sun rose that instant far enough above the horizon to shine directly into the tower's upper dome-like room, and I was awe struck by the texture that the lights created on the glass of the walls, for when it shone through at just the right height, a previously invisible picture came to view. It was of a towering clipper ship with sails that stretched across their masts like skin over the bones of a pleasantly plump fellow, the wind billowing them about at a leisurely rate. Waves broke gently upon the ship's side as the crew rested peacefully on the various cables and nets, all except for the one-legged captain who was busy looking at the map and accompanying charts. It was a quaint and beautiful scene, though it soon passed away as the sun moved upwards in the sky, and I wouldn't have mentioned it, except that as it disappeared, I found myself looking at where it had been, but instead of the ship, I saw directly through the glass the inhabitants of Nunami arising and beginning their daily business, a scene which I might have missed since I was previously wholly absorbed by the picturesqueness of the sky.

Usually the Zards would arise before dawn and be about their business, but because of the great flames of the night before, they had no doubt had trouble sleeping, and therefore slept later than usual when they finally did fall into the lands beyond consciousness. They hustled and bustled about the streets of Nunami, each doing their own business, and there was much business to be done in a city in which all provisions are provided internally, with no trade or commerce outside whatsoever. There were merchants and stores still, yet they were not traders but producers, each making their own wares as they sold ones they had already made. Butchers sat in their shops with their blood-stained aprons already donned, cobblers and tailors were busy with the day's repairs and new creations, the milkmen paraded the streets slowly and methodically, somehow getting their products to the citizens before 8 AM. The farmers and herdsmen were also at work in the fields that were spread throughout the city, plowing and sowing, and being joined by those who had just finished distributing the milk.

All was commonplace and normal, I thought, and I was surprised, for the Zards were not at all martially minded, a great contrast to their

Canitaurian brethren. Of course, I had never actually met any of the Canitaurian commoners. It seems to me that the only ones who really are martially minded are the leaders and politicians, everyone else seems to mind their own business, and sometimes I wonder if there would even be any wars if there weren't any governments with the power to wage one. There was a group of Zards by the government center, which was close to my involuntary quarters, and they were leaning over an opening in the aqueduct that ran down into the lake in the southern section of the city, branching off from there into all the various sectors. They were dumping a barrel of a fine, white powder into the water that was running down into the lake, and after the first had been poured in, they added another and another until they had put a good five barrels into the water source. Once they had finished, they took the empty barrels to a large cage that was down the road a bit, inside of a small grove of trees and shrubs. Inside the cage was a multitude of little beetles that crawled around every which way and were evidentially feasting on a large chunk of glowing material. For a moment I was surprised, and wondered what it was they were doing, but then it hit me: they were the delcator beetles that Bernibus had told me of earlier, the ones that absorbed the radioactive material and stabilized it. As I learned later, they had two good uses, one was that they consumed the unstable materials and neutralized them, but the other was that their droppings, when mixed into the water supply, also gave all that consumed them a greater tolerance for nuclear material. It was almost ironic that their whole way of life was dependent on the feces of another life form, but I will refrain from turning it into a metaphor.

The female Zards wore a black headpiece that mostly covered their faces, and at first I found it strange that for all his talk of progress, the King's people still oppressed their women, perhaps there wasn't as much progress as he had boasted, or, more likely, he was unaware that there was no such thing as progress, just different manifestations of oppression. History repeats itself, they say, and indeed it does, both literally and figuratively.

There suddenly arose a great commotion in the square between the Temple and the palace, and as I looked, I was surprised to see that there was a large crowd gathered. In the middle of the square there were two groups of ten Zards facing each other, with a single Zard in between them, and around the outside of the plaza area stood a hundred or so spectators, apparently watching those in the middle. A moment after I started watching, the solitary Zard, the referee as I found out, walked to the edge, and each of the groups walked to one of the opposing sides and then turned about to face the other. The referee let out a loud yell and in a flash, the two teams ran at each other headlong, until converging somewhere in the center of the field. As they met they dived upon one another and pushed and shoved until the left team had isolated one of the right's players, who was the only one on his team wearing an orange jersey. They dived on him and jumped until the whole field was piled high with them, and then they slowly began to disembark. Once all of the opposing team's players were off of the orange shirted Zard, all was silent and still as the referee held his hand aloft and began counting with his fingers. Everyone held their breathe and stood tensely by as they watched. Just before the referee's tenth and final finger was counted, the orange shirted player rose from the ground, amidst the screams of joy from his team and about half of the crowd, apparently their fans. The two teams then returned to their respective sides, and again the referee yelled loudly, signaling them to rush at each other once more, and more of the same ensued, this time it being the other team's orange shirted player to get pounced on. Once again there was a high pile on top of him, and once again, as they crawled off and he was exposed, the referee began to count. Except that this time the orange shirted one never got up. The other team cheered again and so did the other half of the crowd. The referee went to a pole on the sidelines and put up the number '1' on it while a few bystanders picked the Zard up and carried him off the field. They continued to play in this fashion for awhile, going until one team or the other had no longer any players to be jumped upon, but I was too disgusted at their violent nature to watch, and instead walked over to the end table and picked up the telescope, taking back as I did my thoughts about the innocence and gentleness of the common folk.

With the telescope in hand I went over to the eastern side of the room and began to closely inspect the savanna in an attempt to get a bird's eye view of the point of my entrance in Daem. It looked rather the same from above as it did from below, though the smells and sounds were missing, and I found that it was rather bland once the initial excitement, surprise, and respect of its novelty had worn off. Indeed, it was quite too dull for me, even in my state of boredom as a prisoner, though I suppose that that isn't a proper description of my feelings, for I wasn't free from excitement or intriguing events, but rather, I was in the middle of a campaign of new and anticipated things, but simply unable to participate. Stuck in a room 800 feet from the ground with walls of glass that allowed observation of the whole island of Daem, which I assumed to be the only civilization in the world, while great events unfolded around me, of which I was supposed to be the primary actor, was very disconcerting, though I find in retrospect that fate worked so mysteriously in my situation that it is quite puzzling to think about, meaning, of course, my relationship with the doom of humanity as preventer and provoker, as savior and condemner.

My writing of this manuscript may be considered quite a big cheat, as it details my direct involvement with Onan, the Lord of the Past, and the general circumstances of the end of life on earth, for the current age at least, but still I am allowed to write it. Onan told me just a few moments ago that I could write it and tell all that I want, to which I was taken aback. When I asked why he would allow me to break the law of the council of the gods, he replied that there was no rule against a human agent from detailing his involvement in the actions of the divines. It was allowed, he told me, because it would never make a mite of a difference, for even if it were able to survive the bitter ice ages and all the evolutionary periods in this TAB (Temporal Anomaly Box, which I will explain later, since I get ahead of myself and have not told of them yet), and even if it is found by humans, and even if they are capable of understanding the text contained within it, even then they will take no gain from it. I was again taken aback when he said this, for though I know humans to be stubborn and foolish, in general, I would think that they would at least mind the warning when the conditions of its completion came to pass. But he dissuaded me, telling me that my coevals of the next age would no doubt take it as a novel.

At this I took your defense quite personally upon myself, and demanded in as not so humble a tone as would be thought proper, though as I am about to die within the next day or two, I have to admit that I don't give much of a damn for politics or manners. And yet, with all my ardor I was quickly subdued by a curt rebuke by my interlocutors (for Zimri was there as well), which was, quite simply, that you hadn't taken Homer for any more than a creative poet, even after a few thousand years of study, so why should my meager manuscript make such a large impact. At that, I acquiesced to them and admitted that on that end my attempt to save humanity one way or another was contemptible, but I still write, as you see, for the story's sake, and possibly for my own material immortality. But never mind that, for it is high time that I went back to my story.

I was looking through the spyglass at the various areas of Daem where my adventures had so far taken me. After I had examined them all for a few moments, I felt a strange urge to use the telescope to look closely at the mainland that I had seen before, to see what the effects of the Great War had been there. As I turned the telescope's sights toward it, I was at once surprised and flabbergasted at what caught my eye. There were living beings on the mainland, not too far from the coast. And not only that, but they were standing upright, though stooped, as if by weariness and the wiles of life, and they seemed, in general, to resemble humans, not directly, but as much as the Zards and Canitaurs did, and with the effects of the radioactive instability greater on the mainlands, it would seem natural that they would be further removed from normality than those on Daem. The land itself was barren and flat, with sparse vegetation in the forms of small, deformed shrubs and a short, weak looking grass. As I looked closer I saw that there were about six of the strange, stooped humanoids, and they were gathering the fruits of some of the shrubs for consumption. In a few moments they finished their task and began to walk further inland, and I followed their progress with interest until they finally disappeared behind some of the small plateaus that were scattered here and there among the wastelands.

Putting the telescope down, I walked over to the couch and laid down on it, with indignation filling my every move, for I was almost enraged that the Zards and Canitaurs both should fail to tell me, whom they claimed to respect as kinsman redeemer and whose decisions would seal their fate for good or ill, that there were other survivors from the Great Wars. I was also shocked by their selfishness, for while they fought pettily amongst themselves over how they would change their lands for the better, a seemingly important question about past and future, they completely ignored the sufferings of other humanoids, to whom their way of living no doubt seemed like a paradise. But there they were, stuck across the sea on their desolate lands, unable to cross to Daem and enjoy its plentiful resources and luxuries, yet not at all unaware of them, for as they labored in their hopeless ways, they could see Daem shining like a heavenly vision before them, one which they were not able to touch or grasp, but instead one that must infuriate them to no end in their heart, at the knowledge of fate's unfairness and their utter hopelessness and complete poverty, not because of their laziness or their ignorance or anything involving their actions whatsoever, but simply because they had been born on the wrong side of the sea.

At that moment I was embittered against both the Zards and the Canitaurs for their selfishness and their pretensions of morality. There is no morality where one sees another starving and suffering and does not help, when one sees a whole race of people living on a land where nothing but sorrows dwell, but will not let them share the wealth that was given one by no doing of oneself. There is no morality in selfishness, and when I saw those wretched people, I no longer felt like redeeming those on Daem from the impending doom of humanity. Whatever plans they had for me they never told, I sensed, for there was something deeply wrong about the way they looked at me and talked about me, something deeply wrong about the way they patronized me and treated me like a silly child, while I was the one who was to decide their fate. The Canitaurs and the Zards both looked at me with a subtle sense of deceit and ill will, all that is, except Bernibus, which is why our friendship flourished so swiftly. As I laid there with thoughts of Onan and the decision that I was to make, and of all the responsibility that was put upon me involuntarily, as I thought of the conflict of past and future at the neglect of the present, as I thought about the self-obsession and overindulgence that come with wealth, and the desire for still more that accompanies it, I fell to sleep and into a place where no troubles lay, for my long day and night had left in me no energy for dreams.

Chapter 10
Devolution

When I awoke the sun was once more out in its morning glory, at the height it assumes at about the 9 o'clock hour, and the room was warm and cozy because of it, as it shone in through the glass walls. My first sensation upon waking was one of peace and bliss, the feeling experienced when you wake up late to a nice warm resting place, especially so when all the rest of the world is hard at work and you are not. I breathed in the air deeply and contentedly while stretching my arms, legs, and back in a most relieving fashion, and then turned towards the table in the center of the room, from whence I smelled an extremely appealing smell, that of a hearty breakfast.

As I did so, however, my joy was sent to a bitter, premature death, for there sitting at the table and smiling sardonically at me was the King, arrayed in all his pomp and splendor with his powerful pose, which, while it had impressed, and even to a point overwhelmed me, before, did no such thing to me now, for I was fresh with indignation at the exclusion of the humanoids across the sea from the paradise of Daem.

He saluted me in a polite manner, and I him, though there was little affection behind it. Then, without any more ceremony, I sat down and began to eat, repulsing any attempt of his to start a conversation with persistent vigor, until I had finished, when I stood and demanded where exactly I was to make my toiletry. He laughed and said that he was wondering how long I would last, but as I was still too unpleasant to respond with any familiarity, he showed me to a little room that was tucked off of the side of the bell that formed the entrance to the domed chambers of the upper tower. The top of the tower itself was a half complete sphere, while the room only occupied the upper half, so that the bottom was divided between the entry way and the toiletry room. I spent a few moments grooming and washing myself and preparing for the day, and then rejoined him in the room. He was still sitting on his chair and I took the other. The meal had been carried away.

He began the conversation by saying, "My dear Jehu, I must apologize for keeping you in this position, but you must understand that the outcome of this war is very serious, and I will not risk it to your sensationalism."

"Sensationalism!" returned I, "Is that how you would describe a touch of humanity?"

"What do you mean?" he questioned, apparently interested in what I said.

"Well," I began, regaining myself, my former indignation being exhausted by the spirit of my opening comments, and my normal sober reasoning returning, "I have been observing your society, which you suppose to be enlightened, but I have seen some things, which, I am afraid, are evidences of the opposite."

"Go on,"

"For one, your common folk engage in the most violent entertainment. I saw a vicious game being played not far from here, in the plaza below. There were two sides, and they rushed at each other in a rage and clashed when they met until one side tackled the other. This went on for some time, the evident point of the sport being to gain points by making it so that one of the opposing players cannot get up at the end of a round. It was so brutal that I was disgusted and could watch no more."

"Yes, I see what you mean," the King replied, "I myself would much rather that such games would be forsaken, but the people really enjoy it. I must remind you, as well, that your society had the same type of thing, as did every other before it. It was football for you, gladiators for the Romans, and so forth."

"But I thought that you had no traditions? That you were more enlightened than those of the past? You can hardly excuse your misconduct by reminding one of the misconduct of another, especially when you claim to disclaim the errors of history, or at least, that altered and redefined thing that you call history."

"You are right, I have to admit," he conceded, "But let me remind you that it is a static characteristic of humanity to confuse the ends with the means. When an intense effort is applied, the melodramatic tendency is to honor that effort, despite its uselessness, instead of honoring the product of the effort rather than the effort itself. But, you are right, I admit, for we have still a few places left to refine in the common folk."

Feeling vainglorious at my victory, I pursued him further, "I also observed that your womenfolk wear face coverings in public, which is most certainly a thing of the past."

"I must disagree with you there Jehu," he said, evidentially regaining his confidence and sense of moral footing, "For even in your own time the womenfolk all wore masks and face coverings."

I was taken aback and cried, "Most certainly they did not, your history books may say so, but I, dear sir, was alive and would know best!"

"What, then," he coolly replied, with a sharp grin that reeked of self-confidence, "Would you call all the messes of make-up and perfume and other such things which they were virtually forced to wear? I see nothing different between wearing face coverings and transplanting an entirely new face, hair, and body on oneself everyday. In fact, our women got together and decided voluntarily to do so, for the very reason that if an artificial covering must be put on, it might as well be one that is easy, for why spend an hour or more a day to change one's appearance, when it can be done in moments with a head covering? That is a great time saver for us. And why spend the resources to research, produce, and market massive amounts of facial paint to cover up the face when it is possible to put a covering on and get the same effect much, much easier? It is only logical.

"And in general, Jehu," he pursued, warming to the subject matter, "I find the oppression of women in your time to be quite appalling. You seemed to think that the liberation of women consisted in transforming them into loveless, materialistic thugs, into workaholics whose only desire is wealth, into aggression driven beings that possessed little shred of real humanity, into, in a word, men. I think it would have been a much better endeavor to have attempted to change men into women."

I was taken aback by his eloquent defense of the treatment of women in his society, and felt, I must admit, a little impressed by his arguments, seeing as how it did make more sense to wear a head covering than to paint on a face every morning. Still, I desired to let him see that traditions aren't all that bad, just as they aren't all that good, and, as I had still won one point out of two so far, I felt it safe to move on to my main argument against his humanistic preponderance.

"You are right there, I admit, but tell me, your majesty," I said with a slow, scoffing voice, meant to show that I had a powerful point to make, and as if I had to go slow enough for him to comprehend the eloquence of my speech, "Why, if you are so enlightened and progressive, so humanitarian and merciful, why do you keep a whole race of people, of human beings, stranded on the far shore, able to see the goodness of Daem's plush lands, but unable to visit them? How can you justify the keeping of people in such conditions when it is in your power to relieve them?"

He sobered up more than he already was and answered in his most dignified voice, one calculated to stop opposition by its very graces, "Their plight is unfortunate, but as they are not my subjects, it is none of my concern."

"So you knew of them, but did not care. How typical of powerful men. What are they called?"

"Munams," he answered, "Is what we call them, though people of your time had a different name for them, Neanderthal, if I am correct."

My intrigue superseded my conviction and I asked interestedly, "But, how is that possible? The Neanderthals were the ancestors of men in my time, and the men of my time were the ancestors of the men of this time, how could they be living now?"

"Very simply, for your scientists and philosophers did not understand the revolution of time, and what they thought was evolution was in fact devolution. You see, when they found all the fossils and other such evidence for evolution, they interpreted it to mean that they had evolved from lesser organisms. Since they didn't know that time repeats itself over and over again, ages of time being like the years of the earth, it was actually the remains of the age before them that they thought were the remains of their ancestors. In truth, instead of a great comet hitting the earth and destroying the dinosaurs and many other living beings, it was the Great Wars, the nuclear wars, that caused all the damage. And since their perception of the events was backward, instead of the blasts destroying the dinosaurs and the wholly mammoths, it was what actually created them, for, you see, after the nuclear weapons had all been used, everything in the world died, or came very close to it, all that is, except Daem, which thrived, because of the delcator beetles.

"There were no 'dinosaurs', only Zards, for when the radiation levels were still high and unstable, we grew to enormous sizes, and likewise there were no wholly mammoths, but Canitaurs. And the Neanderthals that appeared shortly after were not the precursors to humans at all, but the Munams, who survived on the mainland near Daem because of the corrected atmosphere, but who were mutilated more than we by the increased corruption across the sea. The Ice Ages, also, were not as you thought, but instead mark the position in the last age after the doom of humanity was played out and everything destroyed. The Big Bang, also, was not at the beginning, but at the very end, being somehow related to the onset of the Ice Ages. Your evolutionary theories were close, but the time tables were rearranged to fit the facts, since time was thought to be linear.

"That is where our main trouble lies, Jehu, for through geological and biological evidences, even more advanced than those collected during your times, we can tell that something happens at this very period of history that will wipe all life from the face of the earth for a long period of time, many thousands of years, until somehow they start to reproduce and grow once more into what they are now. Something very powerful happens, even more devastating than the nuclear wars, when all the nations of the world used their entire stock of weapons. Our problem is how to prevent it, and a great problem it presents, indeed. You see, while we would wish to be confident of success, since we know generally what to expect, we know through research that there have been many, many ages before us in which the same thing has happened. That is why the geological layers have always been found to be strangely misaligned, with fossils from an earlier period here and with a later period there. That is why things like tree fossils are found in coal mines, where they shouldn't be, and why in general, the evidence found in the ground doesn't fit a consistent pattern."

As he finished, I could say nothing, for his revelation was sobering to me, bringing me suddenly back to the realization that our doom was impending, that every decision I made had the potential to either bring us to safety, or to supply the necessary force to hurl us viscously off the cliff of mortality. He was silent as well and allowed me a few moments of meditation to turn his speech in my mind. As is my tendency, I looked abstractly out the window as I thought, fixing my subconscious focus on the road that ran from the northern gate down through the city, the road which formed half of the plaza beneath the temple. A moment or two passed like a solemn parade of mourning, then, suddenly, or at least quite unexpected by myself, a party of Canitaurs came walking down the northern road, unharassed and unescorted through the heart of the city. Since they came freely, I knew that they were not prisoners, but still I was perplexed at how a party of them came to be allowed in Nunami at all under such pretexts, especially as they had attempted to bring it to ruin but a few days before.

The King saw their coming and my interest in them, and said in a way of explanation, "There is to be a council today between the Zards and Canitaurs, with you present, of course. Our war has rampaged for quite some time, but we are forced to peace in light of our impending doom, brought by circumstances outside of ourselves. We will decide tonight, or tomorrow, what action to take. It is a grim time, you can be sure, my dear Jehu, when Zards and Canitaurs meet in peace, a grim time indeed."

He said that very importantly, with an air of fright in his voice, as one who knows his end is near, for both him and his loved ones. There was another moment of silence as he reflected on the meaning of his words, and then he rose and beckoned me to follow him. We made our way through the bottom half of the room and down the long flight of stairs that wound down the great tower in the Temple of Time. When we reached the bottom, we went again into the long room with the bookshelves, the table, and the altar to Temis. Already there waiting for us were the Canitaur emissaries, Wagner and Bernibus.

They rose to greet me, bowing low in a deferential manner, more out of forced respect than awe, at least on Wagner's part, and after the customary blessing that followed, we all sat down at the long wooden table that stretched lengthwise through the room. Wagner and Bernibus took their chairs on one side and the King and myself on the other, he and Wagner being opposite each other, and Bernibus and me being the same; the King and I were facing the altar and the White Eagle that held it.

There was a moment of silence as we took our seats, and it continued for another moment as everyone sat in an awkward situation. As there was no one else in the room besides the four of us, and as Wagner seemed disinclined to begin, the King opened up our conference with the following statement:

"Well, dear sirs, what can I say, except that I am glad that you have finally condescended to seek a mutual agreement on the actions which are about to ensue, and that I hope that our conference will be productive and informative. Before we begin, I will outline the rules of the debate and of the conference, which were agreed upon before the military action of the recent past," here he looked at Wagner with the look of a judge who supposes himself morally superior to the criminal in his holding, "And by which we will still govern the council, despite the sudden change in circumstances. The rules are as follows: The decision shall be made by the votes of the three parties involved, namely the Zards, the Canitaurs, and Jehu, the kinsman redeemer. A majority of two votes is required to decide which of the paths will be taken: the Futurist or the Pastite. As is clearly obvious, my dear Jehu, I shall vote Futurist, and Wagner shall vote Pastite, and it is up to you to cast the decisive vote. You are the kinsman redeemer, and for all intents and purposes, you will be the sole decider of the fate of humanity. It is a great responsibility, but one that you were chosen for by the child of Temis, the God of Time. Wagner and myself will each make our cases, though you know them by now, and then you will have all night to decide and you will tell us your decision in the morning," thus concluded the King's opening address.

Before anyone else could follow it up, I interjected, "But I was sent by Onan to do his work on earth, wouldn't it only make sense for me to choose the way of Onan?"

The King answered me, saying, "You were sent by Temis, the God of Time, Jehu, for Onan and Zimri are his children who do his work for him, but they only have the powers that he gave them. Onan is the only one able to speak to mortals, for he is in the past, while Zimri is in the future, but

Onan also speaks for Zimri, because he is told what to say by Temis, whose agents they both are as much as you are Onan's. Isn't that so, Wagner?"

Wagner sighed in the affirmative, and when he had done so, I asked him pointedly, "Why didn't you tell me? You led me to believe that Onan was the one who sent me, and by his own power."

Here the King put in, "He merely wanted to prejudice you to his own side, Jehu. He attempted to by-pass our peace treaty of long ago when he tried to attack us and capture this very temple for his own plans. We agreed twenty-five years ago to do it this way, because enough blood had been shed, and no good had come from it. He violated it when he took you into hiding, using our pursuit after his treachery as justification. But come, in the face of impending doom we cannot squabble over past wrongs, but must move to prevent future disaster from striking."

"What is so important about this Temple of Time, though?" I asked.

Wagner and the King mumbled together that "It was an essential part of the restoration of Daem", but would not elaborate, saying that it was unimportant to the present troubles. They looked guilty as they said it, though of what I did not know. I was reminded of my indignation at their ignoring of the sufferings of the Munams and became once more impatient with their self-importance, so I yielded the floor and they began to make their cases. In order to decide who went first, they drew lots, and as the shorter was drawn by Wagner, he went first. His speech is as follows:

"The past is constant, Jehu. It has happened and is secure in its place, explored and known. The traditions and customs of our people are steadfast and immovable, for they have survived the ages like a mountain that is untouched by the weather. They have lasted so long not because of the mere namesake of tradition, but because they work, because they have worked thousands of times before, and because we know they will work a thousand times in the future. What was good enough for the generations before us is good enough for us and our children. A tradition, or taboo, is not formed by the decision of some contemporary council as a means to control others via social restrictions, for if it was it would never have lasted, instead it is formed because of experience, because when something goes beyond it the result is temporary pleasure, the nectar of the fruits of rebellion, but when the rebellious desires have faded, what is left is rotten and decayed.

"It brings only more desires for rebellion and more thirst for the forsaking of traditions, and it will not be satisfied. Then another taboo will be broken, but this also will not quench the desires of the rebellious, who do what they do not for any independent purpose, but only from a desire to break traditions and taboos and to be different than their forebears. But there is no satisfaction in rebellion, only in obedience. Obedience not to some alien divinity, not to some social supremest, not to the blind devotion of parental mandates, but obedience to common sense, to practicality, to morality. For a taboo is not formed by any one person, instead it is slowly built up upon the experiences of many, experiences which show that when one thing is done, suffering is what follows, and when another thing is done, happiness is what follows. Of course there are a few, isolated taboos that are based instead on human prejudices, but that doesn't translate into the abandonment of all the experience of precedents. What comes when there are no longer any taboos and traditions to break? Destruction. For as is seen time and again, the rebellion of societies gains momentum, and while their consequences are slow in gathering, in the end they multiply and force those societies over the edge of power, bringing only suffering and ruin.

"And not only are the experiences of the past wielded together into that euphoria that eludes the rebellious - wisdom - but its constant state controls the present and the future. What men have seen in the past leads them in their future actions, and as a result, it is not the future that controls the present and defines the past, but it is the past which controls the present and defines the future. What sense is there in abandoning the mountain of wisdom that the past has built up and leaping blindly into hazy, unknown actions and institutions? The past is steady, Jehu, and it is known; it is the only sensible way." Thus spoke Wagner.

It was then the King's turn, and he said as follows:

"The past is the past, not the present nor the future, its time has been spent, its part in the theater of life is over, it is extinct. Jehu, Wagner speaks of us as rebelliously breaking taboos that were formed by our forefathers, but that is not true. In the present more is known than was known in the past, they had outdated views and opinions, and their ideologies were vulgar and unsophisticated. At present we are more knowledgeable, more refined than what has gone before. The people of the past waged unjust wars. They had superstition and prejudices that clouded their visions of morality, and the product of that is a large amount of taboos and precedents and traditions that are immoral or meaningless. Now is the age of enlightenment, now and never before is the future at hand, mixing with the present as we learn more and more about our world. We are progressive, learning and growing in philosophy and lifestyle.

"If those of the past were so upright and wise, than why are they not still among the living? If they were so powerful, then why are they now extinct? The past is gone, but the future is yet to come, it still holds tangible pleasures, not memories, it has promise and potential, while the past is only the ruins of the same. When the past is looked back upon, it is small and immaterial, it is like time crumpled up into a wad of memories, and a time yesterday or a thousand years ago looks the same, for it is past, it is no more. Life is not short, but in retrospect it seems to be, and its memories are distant, as they float like fish in the oceans of time, lacking both definition and scale, and hanging lifelessly around in random arrays. Every moment is of the same length, but a moment in the past is nothing, its thoughts and emotions are nothing, they are gone and useless to the present, while a moment in the future is long and touchable. A thought that is past is as nothing, and it is forgotten, for the past and the future are like a one-way mirror, you can look forward into the future, but looking into the past you can see only the present reflected back at you. What good are the joys or sorrows of yesterday? They are as far removed as those of a thousand years ago, but it is the joys and sorrows of tomorrow that loom the largest. Why look into the past for completion, when it is found only in the future?" Thus spoke the King.

Once both of them had finished there was a short pause, each reflective and absorbed with his own thoughts. At last the King broke through the still waters of the moment and sent his rippling voice across its formless surface, which revived at once and was joined by many others, until the outward expression of consciousness sent the waters of the mind again into their complex and interwoven dances. He spoke in the department of host and concluded the short session with these words, "Now the cases are stated, though but briefly, for they were already well-known. As planned prior to the infractions of the treaty, we will adjourn for the night, and in the morning Jehu will deliver his verdict, whether we undo our problem through the future, or through the past."

We all rose and Bernibus, my only friend on the island, came up to me and warmly embraced me, while Wagner and the King conversed formally a few yards away. When they were not looking and our backs were turned to them, Bernibus slipped me a piece of paper that was rolled up into a tight scroll. Seeing his caution and secrecy, I quickly stashed it in the inside of my shirt, where it could not be seen. I was alarmed at the momentary expression of his face, which showed that he was greatly worried about me, and made me very interested in what the paper would contain. His face quickly returned to its original countenance, an impermeable barrier to his insides, and no one except myself had any inclination about what had happened. The other two turned towards us, and quickly made their farewells, Wagner and Bernibus departing for their quarters, and the King to escort me back to my prison.

He took my arm in his genially, though only superficially so, for he still had a subdued sense of distrust about him, and we went through the door to the long, circling stairway from whence we had come. As we ascended we engaged in small talk, the usual meaningless pleasantry, which I assume you have probably had enough of in your experiences to allow me to dispense with relating it, for it was of no weight in any of the circumstances that I found myself in, and I especially was not interested in it, as the paper given to me by Bernibus claimed my whole attention, and filled me with an anticipation and mystery of what it might contain. I kept up the small talk with the King merely to allay any suspicions he might have had, though he had none. After a seeming eternity we reached the top, and once there I stepped into my chambers, as the King jestingly called them. We bade each other goodnight, which was followed by the metallic click of the door locking, and the sound his footsteps as he descended and made his way to his palace.

Chapter 11
The Land Across the Sea

I waited reluctantly with my ear against the door until his footsteps could no longer be heard, and then waited for fifteen minutes more, listening carefully for any noises. There were none, and once I had convinced myself that I was completely alone, I dashed swiftly up the stairs and jumped onto the couch. My sudden movements caused the top-heavy tower to sway slightly for a few moments, giving me quite the scare, for I didn't realize what it was at first. But then my pilot's instinct kicked in and I mentally calculated the height and width of the tower and the mass of the dome that rested upon it, and came to the conclusion that it was stable, for while a swift movement caused it to sway, it would take a prolonged and deliberate pendulum-like motion to cause any real damage, and even the fiercest wind would not upset it, for it would only blow in a single direction at a time, and only a rocking motion must be feared.

Confident once more of my safety, I took the rolled piece of paper from the folds of my clothing and opened it carefully. Inside was a note from Bernibus, written in a legible cursive that flowed from an obviously educated hand. It read as follows:

"My Dear Jehu, it is I, Bernibus, your friend and comrade, who writes to you. Wagner and myself are soon to set off for Nunami for a council with the Zards about the resolution of our conflict. It was decided in a cease fire treaty twenty-some years ago that whomever first came upon the kinsman redeemer was to have a council with the other side and the ancient one to decide which course to take, since either course needs the support of both the Zards and the Canitaurs to succeed. When you first came among us, Wagner seemed to break the terms of the treaty and keep you with us in an attempt carry out our plans independently of the Zards, using an attack plan that had been held in readiness since the treaty, to ensure a defense if things went wrong. When the Zards attempted to capture us upon your arrival, Wagner declared the treaty violated, and I assumed that it was to be entirely abandoned. I was under this impression when I befriended you, and once our friendship had strengthened, I had no fears for you, thinking as I did that new methods were to be tried.

"After the attack on Nunami failed and the council was once again to be held, each having violated it equally, my fears were suddenly aroused on your behalf. It was only then that I saw that it was the intention of Wagner not only to destroy Nunami and the Zards, but to capture the Temple of Time, which was the only part of the city to be left intact. When I confronted my brother-in-law about this, he only laughed at me scornfully and told me that I was soft, that I was a fool to put one man's life ahead of the salvation of the whole earth. I was filled with wrath at him and still am, but I have decided that it was better to feign compliance and let you know by letter what it was that is being planned for you. I am only sorry that it should come to you at so late an hour, when I could have warned and helped you before if I had only known. There is not much that you can do now, but still I must warn you, for whatever it is worth, if only to prove my affections.

"You see, my dear Jehu, the Pastites and Futurists interpret the prophecy to mean that the kinsman redeemer has come to renew the earth, as you have no doubt heard, although there is strong evidences to the contrary. I myself have been brought up to this interpretation, as it is more acceptable than the alternate theories that exist, though I have been for a time now doubting its accuracy. According to the Externus Miraculum view, the Temple of Time is crucial to the implementation of either plan, in fact it is the crux of them both, the one issue that it is of as great importance, or greater, than the presence of you, the kinsman redeemer. There is an altar in the center room of the temple, a great diamond White Eagle that is grasping an ordinary altar in its talons, and this altar is where the kinsman redeemer is to be sacrificed. If only I had suspected so before and could have warned when there was yet time!

"But there is no time now for such reflections, so I will continue. The method of sending you back or forward in time is to sacrifice you on the altar of Temis, the God of Time. It is not a traditional, atonement sacrifice, nor of any kind that involves the cutting of the flesh with a knife. Instead it is a molecular one. You are to be set on the altar and then the White Eagle will start to spew forth either protons or electrons, depending on which is chosen, past or future. When your body's cells absorb all of the floating matter, they will be either positively or negatively charged to such an extent that their revolutions will be rapidly accelerated. According to theory, the increased speed of the revolutions would cause a rift in the time continuum, or in other words, would change the proportion between your existence in the temporal and material realms and change your location in time, thereby propelling you into the past or the future, depending upon which was chosen, electron or proton, past or future.

"There has been much experimentation with this process, each person sent through time being equipped with a matter-proof box that is basically an advanced time capsule, lasting for millions of years. Into this box (or TAB, Temporal Anomaly Box) each person was supposed to write an account of their temporal journey and leave it on the island that is presently Daem, at specific locations decided on for that purpose. We would search for those boxes in the present, to see if they had been delivered. None have yet been found, though there are other possible reasons than death, such as a failure to find the island, or the box's removal by someone in an intervening time. Still, I am greatly afraid for your life Jehu, especially so after what I discovered just hours ago in the classified archives of the Canitaurs: there was strong evidence that the process simply disintegrated those upon whom it was tried, instead of sending them through time. This was kept from the public, and was forcefully forgotten by those who knew, their reason being that Temis would guide your travel better than the others who were not called as his servants. If it were anyone but you, Jehu, I would probably have deceived myself in the same way, but I cannot let you be destroyed like this. You must escape and not let them throw away our only chance of salvation in such a way. I only wish that I had known sooner, I only wish that there was a chance that you could escape,

"Your Devoted Friend, Bernibus"

For a moment I could do nothing except sit in silence and ponder over this new revelation. After I had reread the letter twice, so as to be thoroughly familiar with its contents, I ate it, so that if I did escape, or was apprehended doing so, Bernibus would not be found out and suffer because of it, though I doubt not that he would have gladly done so. When I had done that, I ran down to the door and attempted to force it open, but to no avail. Neither could it be picked. And even if it had, it would have done me no good, for there were at least two guards always stationed at the foot of the stairs, and many more between them and the temple entrance, and even if, by some miraculous intervention, I made it that far, that left me stranded conspicuously in the center of Nunami. My only hope was to escape from the island completely, for I would be found soon enough by the cooperating inhabitants if I remained upon their own lands.

The land across the sea then entered my mind, and its degenerate inhabitants, but that was across a wide channel that would be hard to cross even if I had infinite time, freedom, and materials to make a boat which would withstand the waves, and I had none of the three. What little hope I had, then, was out of reach, lost to me like the golden days of the past. It was then that I was overcome by despondency, the hopelessness of my situation weighing my spirits down. It is a peculiar trait of mine that in times of distress and in situations that seem to have no possible favorable outcome I act rashly and without reason. You will remember how I leaned forward and peered into the dark hole when I was stranded on the tiny island in the sea, and how I struck the tree with a limb on the shores of Lake Umquam Renatusum. Likewise, I again did something which would seem illogical and vain: in my frustration, I pushed the table that I happened to be standing against with as much force as I could muster. It slid softly along the carpeting before coming to a halt a few inches from the glass wall. It made no noise or jarring of the floor, but the sudden shifting of weight in the room caused the tower to sway once more, as it had when I had run up the stairs to the couch.

And, as had happened on the previous occasions, the result of my senseless actions was good, as if guided by some external force, for an idea came suddenly to my mind that would not have been there otherwise, an idea that was outlandish and far-fetched, but was at the time my only hope.

I lost no time on preparing my efforts, for there was none to be lost, and set out immediately to remove the carpeting from the floor. Upon examination I found that it was not attached to the ground at all, but only fastened into a wooden frame at the walls that held it tightly in place. It stretched in a circular fashion around the whole of the room and into the center until it came to the stairs that led downward, so that once removed it formed a circle about thirty feet in diameter with a three foot circular hole in its center. In case I haven't mentioned the type of the carpet yet, which I must confess that I cannot remember, I will do so here: it was not a traditional carpet, that form being apparently lost after the great wars, instead it was a silky sheet-like carpet, no more than a quarter inch thick, and in fact greatly resembling the sail of an old clipper ship, the painting on the glass that I saw earlier probably attesting to the fact that it had been designed with that appearance in mind. Like its prototype, the sail, it caught a lot of wind and acted in the same general manner.

Using the bowie knife that was built into the large frontal buckle of the anti-electron suit, which, by the way, I was still entirely wearing, I cut the carpet down its center, making two semi-circular pieces, each with a moon shaped appearance, much like a wing. I based my idea in part on the observation that the Canitaurs and Zards had apparently lost, or disregarded, the springs of my time and instead used a hammock of springy, elastic cords that spread across the face of the furniture. Simply put, they stretched elastic ropes across an empty frame, almost like a trampoline made of individual cords. This created a very comfortable springing feel, for they gave enough bounce to render the surface pliable, but not overly soft. Taking the bowie knife again, I thrust it into the couch, and cut away the cushioning to reveal the support. To my great relief, I found that it was constructed in a manner similar to the other couches that I had seen. There were about two score of the cords, each being between three and four feet long. These I unattached and laid them down in a pile.

Next, I took the four main support beams for the couch, one running along each side and two down the center in a crescent shape, with the same curve and slope as the carpet, as they were designed to contour the same wall. Then I disassembled the table and took from it two of its main beams, which were about a foot shorter than their curved counterparts. These I did not fully remove, instead loosening their screws and swiveling them to extend outwards from the table at a right angle, tightening them again afterwards so that they were secure.

Once that was accomplished, I went to the frame that had held the carpet down and took the pins and fasteners which were used to secure it. These I placed on the crescent beams from the couch, which used the same standard size. Once I had secured the carpet sections to the beams, I attached the couch's beams, via the cords, to the long beams sticking outward from the table, running the ends of all the cords through another cord that could, upon being pulled, adjust their height by pulling or releasing, thus controlling the distance between the upper and the lower beams, and changing the amount of slack in the carpet that was stretched between them. I then removed the legs from the tabletop, leaving just it and the beams together, the carpet being attached to the beams.

Thus my plan was completed, it being, in case you hadn't guessed, a primitive hang glider, the carpet being a sail and the beams the wings, the whole being steerable by either raising or lowering one side or the other, and the altitude being adjustable by raising or lowering the two simultaneously. I felt keen joy at my skills in air travel at that moment, and as I stepped back to admire my work, I felt that peculiar satisfaction of having made something and finding that it was good.

But that moment was short lived, for another problem quickly presented itself, namely, how would I remove the hang-glider from the tower and launch it. It was far too large to go down the stairs and needed to be propelled to a high speed or dropped from a high altitude to become airborne. Since I had no way of propelling it, I needed to launch it from the top of the tower, which provided plenty of altitude, but then the problem of how to remove it from the tower arose. For a moment I was stumped and almost admitted defeat, but then it came to me.

The tower's only weakness was in its lack of protection against a deliberate rocking motion. If I was able to swing it back and forth fast enough by slowly gaining speed and multiplying the momentum, it would be possible to get it to lean far enough that the dome would snap off, leaving the room open to the air. This was possible, though rather unlikely. But I tried anyway.

Starting on one side I began to move from one edge to the other until a faint rocking motion could be felt. Then I increased my speed in proportion to the speed of the tower itself. It was a slow start, but the momentum began to grow, and as it did each successive sway became faster and faster. Soon it was going so fast that I began to have unstable footing, the whole tower creaking like a tree that it is blown by a heavy wind. The speed kept increasing until it reached its fastest, swooshing to and fro with all of its accumulated force.

It was then that the break happened, for on one of the thrusts the top snapped off and the upper dome was flung downwards to the ground. As soon as it was off I shoved the hang-glider with all the force I could muster towards the edge. At first it fell, but a few feet from the edge its wings caught the wind and it was brought up to a stable soar, and just at that instant I landed on it, for I had jumped right after it. I hit with a thud and felt the craft bounce downwards a little as I hit, but it soon regained its stability and sped on through the air as behind me I heard a great crashing sound.

I pulled the left wing down and the glider began to turn in that direction. Since I had launched into the opposite direction of the mainland, I needed to wheel around completely, and as such I held the wing down until I had done an about face towards the east. What I saw was a striking picture: the sun had just begun to rise, and under the influence of its soft textures the city of Nunami looked as it had before: quaint, picturesque, and inviting. But there was a great difference now, for the tower itself had completely collapsed under the momentum, and its ruins had fallen down upon the Temple of Time, demolishing it and leaving only ruins. It had also fallen on a strip of the city, taking with it several buildings and leaving only rubble. The King, Wagner, and Bernibus could just barely be seen amongst the crowds that had dashed out of doors to see what was going on, and I could tell that Bernibus was smiling at my escape as he looked at my wind sailor a thousand feet in the air. A friend who rejoices in your advancement, even at his own cost, is rare indeed.

Turning my gaze upwards, I left Nunami and its troubles behind me and looked ahead to my promised land, and though it was barren and devoid of any significant foliage, it still held something equally dear to me as landscape: safety. The wind currents were strong and my speed was about 30 miles per hour. Great expanses of grassland sped by below me like the memories of yesteryear, and within half an hour I found myself over the ocean.

There is something very refreshing about the sunrise that correlated very well with my present feeling of emancipation, for it is a symbol of the new and fresh, and of the forgetting of the troubles of the past. This was true in my case, at least, for I was soon carefree once more, secure in my freedom. As the wind rushed across my body, I was relaxed in my adopted element, air, though it was slightly difficult to keep myself firmly on the glider, as I was lying unfastened to the tabletop. Below me passed the ocean, looking generally the same as ever, though paler and less alive, like a ghost of its former self, but still close enough to bring the calm of reminiscing.

Soon even the ocean began to give way to the fast approaching mainland, and I abandoned my restive meditations to solve the problem of how to land. I had not made any contraptions for that purpose, having not thought about it in the hurry to leave my prison. I decided to use a traditional circling approach, in the same way scavenging birds descend on their prey. When I was a mile or so inland, I began to circle about in wide spirals, narrowing them as I drew closer to the ground. In this way I had slowed down enough by the time I made contact with the ground that neither I nor my craft was injured in the landing.

The terrain proved to be as desolate as it had appeared from the distance, for the main vegetation was a weakly sprouting grass that was only a few inches high, though not mowed or chewed down. Every few dozen yards there was a single stunted shrub or small tree, or in some cases a group of the same, and the spaces between these was littered with scattered rocks and occasionally a smaller, flowering plant. The topography of the land was mostly flat, though not in the sense of a plain or savanna, instead it was merely a gentle slope, so that the immediate area seemed flat, but in the distance it was seen to rise considerably. There were also a few small hills that were no more than twenty feet high across their whole length, but in the obtuse slopes of the land, even that seemed to be almost mountainous. Brown was the prevailing color of it all for as far as my eye could see, though I cannot say if that condition prevailed inland further, since I had forgotten the telescope, which would probably have proved a useful tool.

A slight wind blew from seaward, scattering the dry top soil about like a cloud of gnats, though there were very few actual insects, and no animals that I could see. The only sound that I could hear was that of the wind howling gently past my ears. I had landed in a sort of valley, which, though not at all deep, was surrounded on all sides by slight hills that prevented me from getting an extensive look at the landscape beyond. Before making any decisions as to which direction to set off, I decided to climb to the top of one of these hills to ascertain my exact situation, and although I was generally reluctant to start off into unfamiliar territory, I also wanted to put as many miles between me and the coast as possible, in case the Zards and Canitaurs came after me, which was still a cause of great anxiety to me.

As I rounded the top of the hill that was directly east of my landing point, I suddenly came face to face with two small people, gnomes by appearance, one of whom I recognized as being Onan, the Lord of the Past. He greeted me familiarly as 'My Dear Jehu', and introduced me to his partner, who turned out to be Zimri, the Lord of the Future. Onan was dressed the same as when I had last seen him, and Zimri was close in appearance, though his hair was long and his beard short, while Onan's were the opposite. Zimri wore a little blue-green frock that fit rather snuggly but not enough to be considered tight. I started our ensuing dialog by saying this:

"I am more than a little surprised to see you upon such good terms with your rival, Onan," giving Zimri an inquisitive glance as I did. "I had just assumed that you two would be bitter enemies, as your followers on Daem seem to be, but I can tell now that that is not at all the case."

He laughed, as did Zimri, and replied, "We are brothers, and as such there is always a strong rivalry, but at the same time there is the closest bond. There is no real conflict between us, but only a trivial and jovial mock conflict, the kind that means no harm and does none, to those involved, but rubs off on others who are less informed, who take it seriously and have a real conflict."

"What do you mean by that illustration?" I asked.

"Nothing. Nothing at all," he sighed, "I have said too much already, it is against the rules, you know."

"Yes, yes, the rules. Tell me, though, how would you say I am doing so far, am I at least doing fairly?"

"Of course, Jehu, you are doing excellently."

"Is it true about the revolutions of time and matter, then?"

"Yes, in fact, it goes even further than that... Say, Zimri, do you think it is allowable to tell him about the physical and the spiritual realms?"

Zimri said nothing, for he can say nothing, but he did nod his head in the affirmative. Thus sanctioned by his brother, Onan continued to speak, "Well, you know that physical existence is comprised of time and matter, and that both of these are involved in a revolving motion, from the minutest foundations to the largest additions. While they both are revolving within themselves, they are also revolving together, around an enigma which, as other of the centers, is completely devoid of the thing which revolves around it, but is found plentifully in them. In the case of matter, it revolves around a black hole, in which there is not found any matter, but there are places of emptiness inside of the matter, in fact, most of an atom is empty space. In the case of time, it revolves around eternity, an enigma where there is no such thing as time, even as there are certain areas where no time exists in physical existence, such as a book. Likewise, physical existence, which is a combination of time and matter, revolves around a place in which there is no physical existence, namely, the spiritual realm. There is no physical in the spiritual, but there is spiritual in the physical. Physical existence is not whole without the spiritual, which binds it together in such a way that gives it life, the ability to think and reason.

"There is spiritual matter in everything, but it cannot be seen or sensed physically unless it is revealed to one by a force on the spiritual side. Or rather, it cannot be understood unless revealed, for it can always be seen through its effects. By this I mean that it leaves a trace in the physical realm, like a jellyfish that leaves a glowing trail in its wake. When the brain of a human thinks, it is not the actual brain that is thinking, instead it is the spiritual matter that exists in the brain, and this spiritual matter leaves a trail where it goes of electric signals and such. When someone feels a certain emotion, such as love or depression, it is felt in the spiritual realm, but its traces are seen in the physical, such as certain chemicals, but these are not the cause of the emotion, only the effect of them. It is possible, through certain drugs, to induce varying emotions, such as happiness or laughter, but these are not the actual emotions, only their physical counterparts, so that while it appears to be happiness, it is not, like the shadow of a man in a field: his form keeps the light from striking the ground beside him, but the shadow is not him, only the trace of him. Making a shadow like the man does not make the man, only the appearance of the man. While the how of a situation may be inferred through physical means, the why is an entirely spiritual matter, and any attempt to observe life without taking into account the spiritual matter behind it will end in the same result as evolution, as the scientists of your day generally imagined it, but which was, in fact, devolution.

"The laws of the physical realm are called science, such as the fact that energy and matter are neither created or destroyed in any natural or artificial process, or that everything left to itself tends toward disorder, or that life cannot come from non-life by natural or artificial processes. The laws of the spiritual realm are called morality. You have no doubt observed that when one does a certain thing, the end result is always good, and when one does something else, the end result is always bad. That is because there are spiritual laws that govern life, and just as there is gravity on the earth that always pulls things down to it, so there is a spiritual law that whenever someone steals something, the result is suffering for both of the parties involved. Just as it is a physical law that man must have oxygen to live, so it is a spiritual law that when someone murders another the end result is always suffering. Why is this, one may ask, but that is a foolish question, or at least a pointless one, for the law of gravity states that on the earth, all things fall downward towards the center of gravity, there is no reason why, except that it is, for it is observed continually to be the case.

"Since men cannot accept that there is a power over them, they deny it, and in the process they misinterpret the various things of life as physical things, not the spiritual things that they represent. For instance, love: men in many "advanced," that is to say, self-obsessed, civilizations, view it only in its physical materializations, but not in its spiritual context. When they see the results of love, romance especially, they do not understand that the romance is only the fruit of the spiritual essence of love, but instead think that the romance is love. There can be so-called romance on the physical level without its spiritual counterpart, but it is only the shadow of love, which will never fulfill and will never be complete, because, by definition, it is only a mocking of the true force of love. On the other hand, true romance is not, as some would seem to think, a certain action or set of actions, such as the gift of a precious metal or some colorful piece of foliage, instead it is whatever is the result of the spiritual love, for the physical manifestation of the spiritual essence of love is not confined to certain objects or actions, but to any that are sanctioned with its blessings. The daily toil of a poor man shows far more love than a lavish gift from a rich man."

When he had finished, I gave him a big grin and thanked him for his lecture, and then asked him how it was that this did not break the rules, but other things did. To this he replied that it affected my task only indirectly, while the other things were all direct concomitants. Then he asked me if I had any other questions for him, and I replied that I did indeed have one. Which was as follows, "I know that there was a great war directly after my departure from my native temporal zone, and that it was very devastating in its reach and effect, and while I know that the situation was very tense at the time, I was under the impression that it was starting to cool down once more. What was it that set it all off?"

"The disappearance of an American fighter jet off the coast of China," he replied straight-forwardly.

My interest was suddenly aroused, for that was the very section where my squadron was stationed, and anyone who was lost would have been a close friend of mine. "Go on," I told him.

"The Americans claimed that it was shot down by the Chinese, and demanded an official apology. That the Chinese would not do, insisting that they had done no such thing, and instead of the whole situation diffusing, as you thought it would, both sides proceeded to war stubbornly, each thinking itself in the moral superiority. But that is as always."

"Do you have any idea whose ship it was that went down? They were all my comrades," I said.

"Of course I know, Jehu, for it was your plane."

"But how? I wasn't shot down, I crash landed on an island."

"But you came to me and I sent you here, and since your radios went out, they had no idea that you were safely landed."

"Still, they must have found the plane!"

"No, you know perfectly well that those islands are brought above and below sea level at different times. After you left, the island was brought below the water, and your plane was lost in the sea, no traces were found."

I was confused, "Onan, does that mean that I was the cause of the war?"

"From a certain point of view, yes."

He was about to say something else to me when we saw in the distance a group of about ten Munams coming toward us, being at that time a few miles away. He then told me that he must leave me again for the present, as he could not interfere directly with my mission. They bid me goodbye and I did the same to them, and then they walked down the opposite side of the hill that the Munams were approaching from. As they walked, they slowly disappeared, until they were gone without a trace, for even their footprints had faded to nothing.

During the time between Onan and Zimri's departure and the Munam's arrival, I was left to myself for a period of inward meditation, an activity that you have probably concluded that I am often given to, which is entirely the case. This new revelation was very troubling to me, that somehow I was the very cause of the destruction of humanity during the great wars, while also the kinsman redeemer over 500 years later, who was prophesied to be the one to bring humanity back into balance with nature, or to thrust it forever off the edge of existence into the damnation of the ice ages. As I told you in the beginning, I am written in the pages of history as the destroyer of humanity, though if it is just or not, I am not able to judge. The name of Jehu will forever be a ripple on the surface of the waters of life, and when it is heard or spoken, the only feeling that it will bring will be hatred and disgust. If only mortals could see below the surface of the waters of life, for just as the ocean can be deceiving on its surface, so can life. Time is like an ocean, but when one looks upon it, what often happens is that all one sees is the present reflected back in its surface, and the eyes are shielded from what lies below, focusing instead on the surface, which is so trivial compared to the abyss which supports it. When one only sees the surface reflected back, then history and its wisdom lose their meaning, and one sees not the past but only the present. What I mean is this: if you look to the past to justify your actions rather than to guide them, you will not see the truths contained therein, but only what your presuppositions already were before you looked, and your ignorance will be reinforced rather than repudiated. Wisdom is the ability to see the past separate from the present, but when one sees the destruction of humanity, he will see only me, his vision being shielded from the true cause of it all, history.

The actions or inactions of one solitary soul cannot bring the end of life, only the accumulation of the wrongs and injustices of a whole race, the human race. Forever I will be eyed as the assassin of humanity, and yet that is not the truth at all, for I am the father of humanity, I am the beginning as well as the end. If you view me only as one or the other, you do not see me at all, but only a pale shadow of my true self. I am Jehu, past, present, and future, I am the concentration of humanity in all its forms and reproductions, I am the creator and destroyer of every age of this temporal maze. Why am I the defender and executioner of the race of men? Why am I the protagonist and antagonist of humanity? Why am I the father and the son, the beginning and the end? Such a question is futile to ask in the physical realm, for here there are no answers to the why's, they are only to be found in the spiritual realm. The physical realm is left only with the how's, and it is those which I am attempting to clarify.

Chapter 12
The White Eagle

It was only a few moments after Onan and Zimri left me that the Munams arrived, for they had run, spurred on, apparently, by their great desire to meet me. In appearance they were like I had seen from afar: hairy and stooped, almost using their arms as legs, but not entirely. Their skulls were large and oddly shaped and their mouths were pushed out from their faces like an ape's. A limp, furry tail hung down from their lower backs, and their hands had a tough, leathery appearance.

There were eight of them, and when they drew near, the foremost hailed me with an eager gleam in his eyes, like one who has long hoped and long been denied. His voice was low and gravelly, but not at all uncivilized sounding, as one would have expected by his appearance, and his facial expressions were equally as livid and distinctly humanoid. He began:

"Hail, the White Eagle, sent by the gods to deliver us! Hail the redemption from paradise, coming to bring us home." With that he held out his arms and embraced me in a very warm, heartfelt manner.

"Hello," I replied, somewhat embarrassed by my lack of authority.

"I am Ramma, leader of the Munams," he told me, "And I welcome you in the name of us all."

"Greetings, Ramma," I replied, "I am Jehu."

"We are joyous at your arrival, oh Jehu of the White Eagle."

When he said this I had a flashback, a moment of memorial deja vu, when the present and the past are morphed together by one thought, when one idea from the past and the present exists in such a way as to connect the two times around it, forming a nexus between the two moments. I was brought back to two separate times, the first being my initial meeting with Onan, when I saw the muraled dome, the genetics of history, and its depiction of the events which were symbolically representative of Daem: the deformed man, the warring races, the worshipers of the White Eagle. The other was my arrival in the Temple of Time, when the King showed me the altar to Temis, the God of Time, depicted as a great White Eagle, wrought in diamond and grasping the altar in its talons. There was something about the White Eagle that connected itself to me inseparably, something that converged us into one form. I had a sense that it was somehow a key to the mystery of the end times, but I could not make the connection. I thought back to what Onan had said to me just a few moments before, that he and Zimri were close friends, and not enemies at all, while those on earth believed their rivalry was a serious conflict. Yet while I had two separate memorial deja vu's, I could not make the connection between them to figure out what they meant.

"Tell me," I asked of Ramma, "What do you mean when you call me the White Eagle?"

"The prophecy said that our kinsman redeemer, who would bring us out of the lands of desolation and into paradise, who would come to us like a giant eagle, soaring high above the sea. Across the ocean there," he said, pointing to Daem, "Is Daem, the paradise land, wherein dwell our enemies the Zards and Canitaurs. They keep us off of the island and on the mainland by force, and here we have suffered ever since the great wars, in these desolate and barren wastelands, where there is neither life nor death, but only a hazy in between. An ancient one with wings like an eagle was to come and rescue us, the White Eagle, and under his guidance we are to be led to victory against our enemies.

"To them he would be sent first, humbly he would come to redeem them from the woes of their own causing, but they would receive him not. Instead they cast him away, and he was to come to us, to bring us to the promised land. What a blessed sight it was when we saw you soaring through the sky on your white wings, and now you have come, my dear Jehu, you have come at last, in the hour of our greatest need. Come, oh White Eagle, and let us go to Kalr, our city. Tonight is the Feast of the Hershonites, celebrating the night that the prophecy was received, and on the same day shall it be fulfilled!"

With that he turned and set off with a step of exuberance to the northwest, the other Munams and myself following him. He walked quickly, and it was all that I could do to match his pace, so that I was left without breath enough to ask any more questions. From what I saw on our journey, the landscape was the same across the whole mainland that was near to the coast, and there was neither change enough nor any landmark conspicuous enough for me to take any bearings. Without the Munam's company, I would have been lost.

Ramma led us on a straight course for about half an hour, there being nothing to steer around, and when that time had elapsed, we found ourselves in a small, battered city. There were no great buildings or infrastructure like in Nunami, nor any complex labyrinths like the Canitaur's military base. Instead there were only weak, unsound huts, built with a framework of oddly shaped driftwood and covered with a thick layer of insulating sod. A road ran through the center of the city, only distinguishable because it was packed down by constant use, and on either side were groupings of the huts in semi-circular patterns, with no space between them left unfilled by soil. This created a wind barrier, preventing the strong winds that whipped across the desert lands from harassing the inhabitants as they worked and played in their communal yards. Each such grouping had a field of a strange, potato-like plant that spread across the back ends of the houses, where the fierce winds piled up loads of nutrient rich top soil from miles and miles around. In the center of the protected areas, each of the communities, for such they were called, had a well that reached hundreds of feet downwards, bringing them almost unlimited supplies of fresh water. Using these two major systems, they were able to live in a comfortable manner, not comfortable in a sense of comparison with the Zards or Canitaurs, but comfortable in the sense that they had food to eat, clothes to wear, and shelter to protect them. Under such conditions humanity can thrive, for happiness is not found in the accumulation of excess comforts, but in the accumulation of excess love. This the Munams had plenty of, and from that point of view were more the evolutionary form of humanity than the devolutionary.

The Munams all wore a sort of close fitting frock, a plain colored one piece suit that displayed their practicality and modesty. It is a hobby of mine to observe the clothing worn by different groups of people and compare it to their characteristics. As I have said before, clothes do not make the man, but the man certainly makes the clothes, and it is possible to judge a person's character by the type of attire that they wear, in that it is an expression of their tastes. The Munams were shown by their clothing to be a very friendly people, for their frocks were hung gently about the body in a manner that was at once both carefree and conservative. This is perfectly analogous to their personalities.

When we came down through the center street, which was really the whole city, for there were no other roads, the people rushed out to meet us, and when they were told that it was the White Eagle, they began to dance joyously about in the streets. There was laughter and play going on all at once, and it was like a great burden lifted from my heart to see them rejoicing, for it almost reconciled their sufferings with the Zard's and Canitaur's ease of life, in that they seemed to be much more happy, in spite of the circumstances.

Ramma gave a short speech to the people, in which he detailed the prophecy and its fulfillment and, in general, encouraged everyone to hope for what was to come. When it was over, he and I retired to his home, which was rather larger than the others and formed its own semi-circle, containing as it did both his private quarters and the official offices of the government, which, while extremely limited in number, were well outfitted. The door of this building opened into a short hallway that had several doors adjacent to it. He led me down one of these and it proved to be a dining hall, though it was not as commodious as most, with only a round wooden table with a few chairs around it and some cupboards and cabinets.

Pulling my chair out for me to sit in, Ramma went through all the normal duties of host with great ease, and within a few moments we were eating heartily from a great dish of boiled potatoes that had been brought in by a servant, or rather, a deputy minister of state, for such was his title. We did little talking before we ate, because I was greatly famished and as such was ill-inclined to be jovial, not that I was sullen, but I found it hard to be completely relaxed without a full stomach. Yet when that was remedied and I found myself satisfied and comfortable in a warm dwelling, I opened up to Ramma and we had a long and entertaining discussion, some of which I will record here, as it shines a little more light upon the mysteries of my story:

"So, my dear Jehu," Ramma began, "I trust your stay on Daem has so far been enjoyable."

I chuckled quietly and told him, "No, not entirely, for there is a war afoot on Daem, or at least there seemed to be, and it made quite a bit of trouble for me."

"I'm sorry to hear that," he replied, "But also gratified, for it will help us in our offensive if they are against each other as well as us. Still, it will be hard."

"What offensive is that?" I asked, my interest being perked.

"Our jihad, to capture the lands which were meant for us and reclaim them from the filth that now inhabit them. You are our kinsman redeemer, Jehu, but it is not with your presence alone that we will be brought victory, for we also must act. Ever since the prophecy was given we have been preparing for a strike that will catch the Zards and Canitaurs by surprise, for those are our only advantages: time and surprise. The carrying out of the surprise attack is the hardest part, and we decided long ago to dig a tunnel under the sea to bridge Daem and the mainland, for if we had made a fleet of ships, or attempted anything on the surface, they would have seen and known what we intended to do. The tunnel is very long, and it was an arduous task to undertake, but with much patience we prevailed, and now it is complete. In fact, it was only completed yesterday, though it was started more than 500 years ago."

"How is it that you started so long ago and only finished just before I arrived? I asked.

"Fate," he answered, "All the happenings of the world are controlled by a force much greater than us, and it brings everything into completion when it is needed, no sooner and no later. Many civilizations try to out wit fate, but they cannot, and in the end they do its bidding. Not, however, in the way they had planned, and with more consequences than they would like, at which point they try to change fate again and undo those consequences, and soon they are in a downward spiral of such deeds. We recognize that we are controlled by fate, and instead of fighting it, we go along with it. We know that things will happen as they are meant to happen, and we knew that 500 years ago, so it was no great trial for us to work at our task for so long and not to know when things would be brought to completion. You see, if we had worried about it and attempted to change to course of events that history dictated, than we would have only given ourselves more work for the same end. Stress is the only thing that is created when you try to alter fate, so it is our philosophy to take things as they come and trust to the powers that be. You may think it unsophisticated, but that is just as well, for what matters is not appearances, but reality, and we have the two things that matter most in life: peace and joy."

I agreed with him, for I had found the same to be true in my own experiences. I then asked him, "When will this grand offensive be undertaken?"

"Tomorrow," he said bluntly.

"Tomorrow? Isn't that rather soon?"

"Why? Fate has been fulfilled so far, why wait when it is time to act? Maybe you misunderstood my meaning: it is not our philosophy to simply let things go as they will. Instead we relax and let things take their course when it is not in our power to do anything effective, but when the time comes to act, we act swiftly and do not delay. In a word, we do not force fate, either by forcing action where patience is needed, nor by forcing patience where action is needed."

"That sounds well enough," I said, "But the difficulty lies in the correct classification of the situation, or in other words, deciding if patience or action is needed."

"Yes, of course, but in this case it has been decided to attack tomorrow, and there is nothing left to do but to attack tomorrow. But do not yet let your spirits be dampened by the onset of war, for tonight is the Feast of the Hershonites, and there will be great celebrating and rejoicing this evening. Forget about the troubles of tomorrow and enjoy the celebrations of today, as I always say. And it is now time for the celebrating to begin, so let us be off."

And with that we both rose and took our plates into the kitchen that was connected to the dining hall on the opposite side as the hallway and deposited our plates to be cleaned later (for even the leaders of a society must do their fair share of the work). Then we walked back through the dining hall, down the hallway, and out the door.

Outside we found that the people had already began to assemble on the road in front of their communities and were preparing for the festival by chattering with one another as loudly as one would think possible. A hush began to fall upon them like a descending fog when we came out, though, and within a few moments it had died down to a ghostly silence, for all that could be heard was the wind's constant blowing. Ramma took the head of the procession of Munams that had formed on the road, and I took the place next to him. With a sort of quiet anticipation of the joys to come, there was little movement, and what little there was, was hushed by a sense of subdued excitement. Then, with a somber gait, Ramma began the parade down the road, in the opposite direction as we had come from, that being northwest, and all followed him as he did.

The sun at that time was just beginning to set, and once we had crossed one of the larger hills we came face to face with the coast, the sun's great red form half sunken beneath its surface. A faint cloud layer floated by and was illuminated by the twilight so that it stretched haphazardly across the face of the sun. Never have I seen so profound a scene as that which then presented itself, with the desert sands and the ocean's still surface reflecting the last agonies of the sun's descent into the underworld with such a subtle emotional undertone so as to render it a subconscious delight. Its recognized superiority to mortal life forms left us all mute and somber, but at the same time the freedom felt from the same gave us joy beyond reckoning.

The march to the sea was slow and steady, and when we finally reached its shores it was just at the change of day and night. Several large bonfires were lit and by their light a great communal dance began, everyone jumping around, running, and doing whatever their lighthearted desire may have been. Under stars that shone like the twinkling in a newborn's eye, we had such a joyous time that it can hardly be described. We were no longer within the reach of civility or social duty, but without it we were not mean nor hurtful to one another, but were playful and joyous, like children without a care in the world. Our little games and frolics cannot be described with any accuracy, because outside of the moment's happiness, they cannot be understood, as it was a spiritual happiness, existing only in the spiritual realm. All that could be described is the physical actions that were taken because of that spiritual enjoyment, but that would do nothing to describe the feeling of the night. It was one filled with more joy than anything I have known as an adult, because we became as children in our trusting to fate, and it was natural, befitting to our natures. Man is not meant to worry, man is meant to be free from all boundaries, inward and outward, man is meant to be ruled by only one desire: love of others.

As the night dwindled away, we grew tired, but instead of returning to the city, we laid down wherever we were when we felt that we could remain awake no longer, and fell to sleep instantly when we did. It was not at all uncomfortable, for the sand was soft and a warm breeze blew in from the water, and though as an adult I would have feared sleeping so openly in the unknown, I was not at that time an adult.

Chapter 13
The Big Bang

The Munams and I were all awoken at the same time late the next morning by a loud trumpet blast that shook the very air around us with its intense bass. For the first moment of our consciousness we were all dazed and could not fully comprehend the situation, and for a brief time we all sat unsteadily around the beach where we had fallen asleep. As we grew more awake, we began to understand what had happened, or at least I did, and I was frightened when I looked around and saw where the trumpet blast had come from: the entire Zardovian and Canitaurian armies were assembled around us, having somehow crossed over to the mainland in the night, while we slept peacefully, unaware of their presence.

My first thought was for myself, and what would become of me in the wrath brought on by my escape, but that soon vanished when I thought of the Munams, for they were the enemies of those on Daem, even more so than those on Daem were to each other. We were completely surrounded, with the ocean on one side and the Zards and Canitaurs circling us in the front, the former on the left and the latter on the right. All of them were equipped for war, with swords, spears, and shields held firmly in their hands, and thick, leather armor stretched across their chests. The Canitaurs had especially come prepared, for they had brought all of their atomic anionizers with them, enough combined fire power to level the entire world several times over.

Within five minutes, all of the Munams had assembled behind me and Ramma, who stood between them and the Daemians. They huddled closely together and quaked slightly in fear, for they evidently thought that their plans had been discovered and their enemies had come for revenge. I, myself, thought that they had come for me, and Ramma's opinion could not be guessed, for he was a statesman first and foremost, and when his people were in need he rose to the occasion with all the power and grace allotted to mortal beings.

Wagner and Bernibus broke the Canitaur's ranks and drew near to us in the center, as did the King from the Zard's. They reached us in silence, and for a long moment there was no talking, for all present knew that something grave was about to happen, something that would decide the fate of the men of this age, whether they would pass or fail the test. Bernibus looked at me with entreating eyes, showing his sorrow at my recapture and asking for forgiveness, but I had none to give him, for he had done no wrong to need it. He had no power among the Canitaurs, but was only a titled commoner, more like Wagner's groom than counsel.

I noticed that the Canitaurs were not wearing their anti-electron suits, which was strange, for they had brought a few hundred atomic anionizers, though I didn't question them about it, for the answer was evident enough when I had given it some thought: the Zards had no such suits, and were afraid that the Canitaurs would destroy them and Munams at the same time, for while they were allies against foreigners, they still did not trust each other. I still wore my suit given me for the raid on Nunami, though I had forgotten about it due to its comfort. That made me the only person on the earth still wearing one, the only one safe from the anionizers.

It was an overcast morning, and the air was damp with a cold, wet wind that blew in forlornly. The ocean's steady swoosh added to the scene, making it as depressing as the night before was joyous, and in the bluish half light all was colorless and hopeless. At length the King spoke, saying, "My dear Jehu, I am very disappointed in you. Not only did you flee from us irresponsibly, but you destroyed the Temple of Time and the altar to Temis. Without the White Eagle, the prophecy says that there is no hope for humanity."

Wagner added, "And now the only way left to bring about the completion of the world once more is to sacrifice you using the old methods." This he said with evident pleasure, no longer feigning to be my friend.

Here Bernibus entered the dialog, throwing away his timidness with one quick motion and saying to Wagner, "You scoundrel! You said that we came to retrieve Jehu, not to sacrifice him. How is it that you lied to me in such a manner?"

"You fool," Wagner said, "If I had had my way, you would have been dead long ago. You have no authority here, so begone."

Bernibus grew angrier, a terrifying state for a Canitaur to be in, and he was a strong and powerful one at that, though his meek nature had hidden it before. "You would never dare to kill me in the open, you coward, the council would banish you," he said.

Here the King joined in once more, laughing, "He wouldn't, no, but I would. Do you really think that we found your outpost on our own, oh Bernibus the 'deputy kibitzer'? You know that we have no tracking ability, and least of all in your own territory."

Bernibus grew more enraged, and the King was spurred on by it.

"Oh yes, you know what I speak of. Your brother-in-law told us where you and your wife were living, and not only that, for he also told us when you would be there."

Bernibus became even more flushed with anger and vehemently asked Wagner, "Why, you heartless brute? What could you possibly value more than your own sister's life?"

"It was a pledge to the Zards of our intention to abide by the agreement, what more precious thing could I give then my own sister?" He spoke calmly and spitefully, enjoying the end of his long charade of nicety, "Besides, the council was falling for her peace talk, as they always give great heed to every member of the royal family, and I was not strong enough at that time to control them, as I do now. Unfortunately for me you were out at the moment of the attack and able to escape, but still it was a favorable outcome," Wagner said, sneering at Bernibus' outrage.

But Bernibus was not to be taken lightly, and neither was he to let the love of his life go undefended. He leapt at Wagner and grabbed the remote to the atomic anionizers from his belt, where it was always clipped. Wagner tried to get it back, but Bernibus was too strong and hurled him to the ground. Then he took a few steps backwards and stood his ground far enough from everyone to have at least a moment to react before they could reach him. He held the remote out towards Wagner, pointing it at him as if it were itself a weapon, with his thumb and forefinger in position to set it off at a moment's notice.

"Bow before me now, Wagner, or I shall destroy us all," he demanded with a grim smile that showed his resolution.

Wagner did as he commanded and fell to his knees in front of Bernibus, saying in the same gentle, appeasing voice that he had first used on me, "My dear Bernibus, do not be rash, do not act in anger. Let's talk this over, and see ... and see if we can't find a peaceful solution," his fear of death evidently caused him to stammer.

"You fool, do you think that I haven't heard that voice a thousand times before? Do you think that I will fall for your same trick once more?"

Wagner put his face to the ground and groveled like the filthy swine that he was, for he knew full well that if Bernibus set off the atomic anionizers he would die. His life was completely out of his hands and there was nothing that he could do to reclaim it, except to beg for forgiveness. This he did, saying, "Bernibus, you do not understand, the situation was more complex than you realize, and I had no choice but to act as I did. Do you not think that it was as hard on me as yourself? She was my sister, my only sibling

But there was no other way, I had to put the advancement of our people over the life of anyone, even my own sister, as you must do now, putting the advancement of our people over petty differences."

Here the King interjected, "Bernibus, do not act rashly, I beg of you, for if you set off the anionizers, than all is lost. Do you not realize that if you do that, all that we have worked for all of our lives is lost?"

It was Bernibus' turn to sneer, and he did, raising the skin above his teeth and scowling fiercely at the King. "What is it that we have worked for all of our lives? Do you still not understand? You and Wagner plot to return the world to its former glory, each by his own way, but take a look around you. The trees on Daem are taller and stronger than any known before, the grasses are thicker and livelier, the waters are purer and cleaner, the wind is fresher. You know no suffering. The prophecy had nothing to do with you, and nothing at all to do with the restoration of the world! Can you not see that what you have is far more than you have need of, that there is no desire left unfilled in your lives, except that of ultimate power? This world does not need to be restored. Only your hearts have need of that.

"The prophecy was given for the Munams, who were left stranded here in this desert wasteland, while across the ocean they could see the great paradise of Daem, the great paradise that you took for granted. There is to be no restoration of Daem to its original form, but a restoration of the Munams to Daem. You struggle to restore Daem, but have no compassion for the suffering of humanity across the sea. You are the fools, not me, and you are the ones who have brought us all to the very brink of destruction, to the ice ages which you have tried so hard to prevent. Do you not see that Daem is already the paradise, that the only thing that it needs for completion is the residence of the Munams? Jehu is not our kinsman redeemer at all, he is theirs." Here Bernibus seemed to lose his anger and passion and become meek once more, saying humbly, "You have destroyed the life of one whom I held more dear than myself, but that is past, and I will not destroy us all for vengeance.

"Zards, Canitaurs, and Munams, hear me now and listen to my words," he continued, speaking to the amassed groups of the armies that had been listening closely to his words, "We are not separate people at all, we are not different races. We are not Zards, or Canitaurs, or Munams, we are Daemians, and it is time that we came together, to help each other instead of hindering. Look at how much blood has been shed, how many lives have been lost, must we all be drowned in the blood of our brothers before we realize that we are one people? Must we suffer more than we already have in an attempt to undo what has already been done? More pain will not negate the pain that has already been felt, it will only result in more suffering than we have known up to this time. My friends, we need not look for our redemption in the past, for it has gone and though it influences us, we are not bound to its suffering. And we need not look for our redemption in the future, for it is not yet here, and when it comes it will only be what we make it. Instead let us look for our redemption in the present, where it can be found, let us put aside our hate and our divisions and become one flesh and blood, one body. People of Daem, let us live in peace!" As he said this, the Zards and the Canitaurs and the Munams all let out a joyous shout of agreement, and there was seen on every face a remnant of the happiness that had so long alluded them in their wars.

To emphasize his point of harmony and trust, Bernibus dropped the remote to the atomic anionizers to the ground. But it would never land. Wagner leapt forward from his groveling position and grabbed for it as it fell, reaching out with all his strength. There was a sudden silence that overtook everyone as they saw what was happening. Bernibus looked down and saw Wagner leap, but he was too late to prevent him from reaching the remote. There was no noise at all, for everyone looked in horror at Wagner's plunging form. As if in slow motion, his hand wrapped around the remote and he squeezed it so as not to let it go. But as he did so, there was a loud beeping sound that came from his fist: he had triggered the anionizers.

The eager faces of everyone there, of everyone alive on the earth, was turned towards Wagner. The remote had a five second delay built into it, and those five seconds were the longest of my life. Bernibus' eyes met mine, and we experienced an intra-personal deja vu, the converging of the presents of two minds. His face showed the depths of his being in that split second, and he was peaceful. Though he was about to be destroyed, he had no fear, no regrets, and in those five seconds, while Wagner and the King were frightened and frantic at their impending doom, Bernibus was as calm as ever. As I looked Bernibus in the eyes, I could hear Wagner break the dead silence with a shrill scream that echoed across the horizon and ripped through the hearts of every hearer. When faced with death he had no courage, no strength to face the unknown beyond the veil that separates life from death.

As I turned and cast my eyes across the horizon, I saw the faces of hundreds of men, whether Zard, Canitaur, or Munam, and written on everyone of them was a great despair, for they stood unprotected in the presence of death. It was like the calm before the storm, those five seconds, and through them time seemed to stop, to be non-existent, and there was not a sound to be heard, except for Wagner's scream. Oh, what anguish was written on the faces of all around, standing defenselessly before the end with neither will nor way to stop its terrible approach, oh, what fear filled their eyes as their mortality was made manifest before them like a vulture's approach, oh, the pain, as fate stood before their distraught faces and silently whispered, "And to dust shalt thou return."

But then even that was silenced. There was no noise. As I looked upon them they were destroyed, before my very eyes they breathed their last and were no more. One moment they were normal and healthy, and the next they disintegrated, falling into little heaps of limp skin and bones. In that moment I felt a horror such as I have never felt before, a complete loneliness, like a night that never ends. There was no one, nothing, around me. The force of the blast had leveled the already flat terrain completely. The ocean was suddenly solidified into the same lifeless, inorganic mass that the land had become. Across the channel, Daem was no more. There were no more trees, no more grasses, no more cities, no more mountains, everything was leveled, decimated. The sky began to turn a dark, bloody red, and the sun was hidden behind it. Like a disease it spread across the horizon, devouring the light hearted blue and leaving only red: lifeless, deathless red. There was no wind, no sound. I was all alone, I alone had survived the blast because of my anti-electron suit. I gazed in absolute horror across the field where only seconds before thousands souls had been congregated. I looked at its emptiness and I saw nothing, for there was nothing. They were all dead. Every single one of them.

Chapter 14
Past and Future

I have no recollection of how long I stood there staring blankly into the void, for the sun was hidden behind the darkened sky. I have no memory of that period until I saw two short forms coming towards me in the distance. They walked slowly and methodically, as if they were not hurried on by any physical concerns. As they drew near, I saw them to be Onan and Zimri, the Lords of Past and Future. When they arrived I was awakened from the trance that I had fallen into, and I gave them a slight bow, for I was still standing upright. The look on their faces was one of sorrow, for no matter how many times they had seen the destruction of humanity, each time it brought only fresh, poignant sorrow.

Onan was the first to speak, breaking the silence with a long, hopeless sigh, "My dear Jehu," he said, "This age has come to a close."

I could say nothing, for Bernibus' face was still gazing at me in my memory.

"Do not be saddened by grief or guilt, Jehu, for it is what has always happened. It is not your fault, for the events that you have witnessed do not have their roots in your time or in this one, but in the very foundation of the world. It is not your actions that caused this, but rather the accumulated momentum of all the ages of humanity, for they are history, and history reigns by influence. There were no right choices and no wrong choices for you, for the power of the kinsman redeemer is not in himself, but in the way that those around him react to what he signifies. In every age before this you have done the same, as you will in every age after this as well. You were humanity's last chance, yet it is not up to you to change their course: it is up to them to change their own."

Here I raised my head from its dull droop and looked questioningly into his eyes. "What do you mean," I asked, "That I did not prevent it in any of the other ages? How could I exist in any other age but this?"

"Then you do not understand?"

"Why else would I ask?" I faintly smiled.

"These are the Ice Ages, the end of an age of history. Every time that the temporal continuum revolves around eternity, it has a new age, much like the years of the earth as it revolves around the sun. When the atomic anionizers went off, they did on a large scale what they were designed to do on a small scale: reverse the poles through an extreme electric charge, by injecting countless solitary electrons into the atoms. But with so many of them exploded at once, they did this to the earth itself, reversing its poles. It was a theory at your time that the poles reversed about every 170,000 years, this is because that is how long an age is.

"When the earth's poles were reversed, it brought all to desolation, excepting you, for you were protected by the suit. But while this is the ending of all life on earth, in a way it is also the beginning, for you see, Jehu, you have just witnessed the Big Bang. In a few days, at the longest, you will die yourself, for there is no food or water for you here, but inside of your anti-electron suit, your remains will be protected. Slowly the earth will regenerate, and when conditions suitable for life have been once more returned, your suit will be blown against a rock somewhere and broken open. From that little hole, the atoms of life, your life, will escape into the atmosphere and grow and evolve until they become like what things were before you were born. Then the process will be repeated. You are not only the one who symbolizes the destruction of humanity, but also the one who symbolizes the rebirth of humanity. You are the beginning and the end, in a sense, a descendant of yourself, simultaneously the father and the son. You will be born again through your own descendants, and will once again become the kinsman redeemer. It is your destiny, there is no other way. You are the White Eagle."

"You only confuse me more, what is this White Eagle?"

"Do you remember when we first met, in the Chambers of History? On the dome of the ceiling there was a sculpture mural, and in it was a White Eagle, holding many lords and ladies in its talons while it soared far above the lands, and those on the land were worshiping it. You are the White Eagle. You hold all of humanity in your hands, for you are the father of all men, they all descend from you, including you, yourself. You were the White Eagle, for the altar had no power, the power was only in you.

"Those who worshiped you were those who worship time, in either of its forms, past or future. Those who worship the past recognize the influence of history, and they understand that there are taboos and traditions created through mutual experience. These traditions reign in humanity by keeping men from actions that lead to pain and suffering. But they do not understand that while it influences mankind, the past does not control them, for it is gone, and it will never come again. In their strict keeping of traditions, they focus on the physical act of the tradition, while neglecting the spiritual principle behind the tradition. If you keep only the physical form of the principle, you have nothing.

"On the other hand, those who worship the future neglect the past and the valuable lessons that it teaches. They believe that there is some moral advancement that places them above those that have come before, they believe that the people of the past were blinded to the truth, and that the revelation of the truth in the present supersedes the traditions of the past. But they are wrong as well, for humanity is humanity, and those of the past were no more ignorant than those at present. The people of the past fell into the same traps as the those in the present, and both suffer the same consequences.

"While one group remembers only the physical display of the spiritual truth, the other rejects the spiritual truth because of its physical display. Those who worship the future break taboos because they recognize that the mere physical manifestation of the truths is not their entire essence, but they reject the spiritual truth as well. When taboos are broken, there is nothing gained, but everything lost, for the physical traditions at least lead to the knowledge of the spiritual laws to those who seek such wisdom. One taboo is broken, but as there is no satisfaction in the breaking of taboos, every one of them is broken in succession. Then there is no limit to the immorality that is left to freely roam the hearts of men, and when immorality, the breaking of the spiritual laws, is widely propagated, there is spiritual suffering. When this spiritual suffering begins to accumulate and is translated into physical suffering, the people see what is happening, how their very society is crumbling to ruin around them. Yet instead of recognizing the truth of what is happening, they see the traditions of the past as the cause of their problems, and continue to make their plight worse. This downward spiral continues until at last we find ourselves where we are now, at the end of an age."

"But what else is there to do?" I asked Onan, 'If both the past and the future lead to ruin?"

"The answer is in the present, my dear Jehu, for if one focuses on the spiritual laws that bring good or evil, and acts according to them, instead of their physical counterparts and manifestations, then things will thrive and become prosperous. What is evil brings evil consequences, and what is good brings good consequences, over time. The ends define the means, just as the fruit shows the tree to be either good or bad. These spiritual laws become known and remembered, not why they are so, but simply that they are so. No one can question why, for morality is observed through its effects, just as science is. When people observe that one thing brings good and another bad, they remember to stay away from the bad things and cling to the good. Over time these evolve into taboos and social restrictions, not meaningless laws enforced by tyrants for their own reasons, but rules that are observed by all because the are the laws of the spiritual realm and govern physical life. But when the people forget what the traditions represent, then all is lost, and either of the two paths that present themselves lead to ruin."

"But why do not men see?"

"Because they are rooted too strongly in the physical realm, and cannot, or will not, see the spiritual. What they see as happiness is not the spiritual matter that is happiness, but the physical actions the represent happiness. What they see as love is not love in the spiritual sense, only its manifestation in the physical realm. When they see the happiness that comes from a spiritual connection, they seek after it. But they do not seek after the actual essence of the spiritual connection, yet after its physical counterpart, marriage. This they take and defile, and when they go through the physical actions of the spiritual marriage but forsake the very thing that makes it bring happiness, they are left without any real sense of satisfaction, without any real happiness.

"You must understand that the physical manifestation of the spiritual force is not the spiritual force at all, only a bland deception. If you only focus on what you can see directly, than you chase after only the representation and not the object desired. If a bird is flying through the sky at noontime, casting a shadow on the ground below him, and a man comes along, and in the hope of catching the bird chases after its shadow, it is evident that he will never catch it, for when he does reach it, he will find that there is nothing there at all, only the shadow of what it was he desired. So it is with the spiritual!"

"Yes, I think that I am beginning to understand."

"Excellent. If only I could tell you more, but I must go, my dear Jehu, for Father Temis is in mourning for his children, and I must go to comfort him."

"I thought that you and Zimri were his children?" I asked.

"You are all his children. He is patient, ever so patient, but still they fall by the wayside, too caught up in their false perception to rest in him. Fare thee well, Jehu, may you be blessed ere you must die."

And with that, Onan and Zimri turned and walked away in the other direction, never to be seen by me again, in this age. I took a look around me, and could not bear to remain any longer in a place of such ill remembrance. Turning slowly and despondently to the westward, I began to walk over the lifeless mass of what had been the ocean not too long ago. For how long I walked, I could not tell, but in due time I reached Daem, though it was no more hospitable than the mainlands, for all was laid to ruin by the Big Bang, all was equally devoid of life.

When I came to what had been the center of the savanna, I came across something that had survived the blast, being unearthed from its previous burial hole by the force of the anionizer's explosion. It was a two foot by two foot box, made of a strange metallic substance with an intricate etching along its top. Written there in its center were these words:

"Temporal Anomaly Box, Number 12, Location: Central Savanna"

I took the lid off carefully, though it was in perfect condition and I did not need to treat it so, and looked inside of it. There was a notebook and a pen there, both capable of producing a large ot amount of enduring text. This was one of the boxes that had been taken back through time in the experiments of the Zards and Canitaurs, designed to withstand any conditions, and to hold its contents for countless ages, until they should be retrieved and studied. I sat down on the ground and began to write my story down, in order to assist whoever takes the job of kinsman redeemer in the next age. I knew that it would have all been forgotten, so I made sure to carefully record it, for it could mean the difference between the life and death of humanity.

This was only hours ago, and now I have reached the end my tale. If by any chance you come upon this in some subsequent age, I beg you to take heed, for what I have written will surely come to pass once more if something is not done to prevent it. There is nothing else for me to say, for this is the end of my story, and within the next day I will also pass over to the spiritual realm. What, then, can I say to bring this to a close, for this is neither the end nor the beginning. I suppose all that can be said is this: